POEMS THAT SOLVE PUZZLES

POEMS THAT SOLVE PUZZLES

The History and Science of Algorithms

Chris Bleakley

OXFORD
UNIVERSITY PRESS

OXFORD
UNIVERSITY PRESS

Great Clarendon Street, Oxford, OX2 6DP,
United Kingdom

Oxford University Press is a department of the University of Oxford.
It furthers the University's objective of excellence in research, scholarship,
and education by publishing worldwide. Oxford is a registered trade mark of
Oxford University Press in the UK and in certain other countries

First Edition published in 2020

Impression: 2

Published in the United States of America by Oxford University Press
198 Madison Avenue, New York, NY 10016, United States of America

British Library Cataloguing in Publication Data
Data available

Library of Congress Control Number: 2020933199

ISBN 978–0–19–885373–2

Printed and bound by
CPI Group (UK) Ltd, Croydon, CR0 4YY

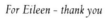

For Eileen – thank you

algorithm, *noun*:

A process or set of rules to be followed in calculations or other problem-solving operations, especially by a computer.

The Arabic source, *al-Kwārizmī* 'the man of Kwārizm' (now Khiva), was a name given to the ninth-century mathematician Abū Ja'far Muhammad ibn Mūsa, author of widely translated works on algebra and arithmetic.

Oxford Dictionary of English, 2010

Foreword

This book is for people that know algorithms are important, but have no idea what they are.

The inspiration for the book came to me while working as Director of Outreach for UCD's School of Computer Science. Over the course of hundreds of discussions with parents and secondary school students, I realized that most people are aware of algorithms, thanks to extensive media coverage of Google, Facebook, and Cambridge Analytica. However, few know what algorithms are, how they work, or where they came from. This book answers those questions.

The book is written for the general reader. No previous knowledge of algorithms or computers is needed. However, even those with a degree in computing will, I think, find the stories herein surprising, entertaining, and enlightening. Readers with a firm grasp of what an algorithm is might like to skip the introduction. My aim is that readers enjoy the book and learn something new along the way.

My apologies to the great many people who were involved in the events described herein, but who are not mentioned by name. Almost every innovation is the product of a team working together, building on the discoveries of their predecessors. To make the book readable as a story, I tend to focus on a small number of key individuals. For more detail, I refer the interested reader to the papers cited in the bibliography.

In places, I favour a good story over mind-numbing completeness. If your favourite algorithm is missing, let me know and I might slip it into a future edition. When describing what an algorithm does, I use the present tense, even for old algorithms. I use plural pronouns in place of gender-specific singular pronouns. All dollar amounts are US dollars.

Many thanks to those that generously gave permission to use of their photographs and quotations. Many thanks also to my assistants in this endeavour: my first editor, Eoin Bleakley; my mentor, Michael Sheridan (author); my wonderful agent, Isabel Atherton; my bibliography wrangler, Conor Bleakley; my ever-patient assistant editor, Katherine Ward; everyone at Oxford University Press; my reviewers, Guénolé Silvestre and Pádraig Cunningham; and, last, but certainly not least, my parents and my wife. Without their help, this book would not have been possible.

Read on and enjoy!
Chris

About the Author

Chris Bleakley has thirty-five years of experience in algorithm design. He has taught and written on the subject for the last sixteen of those years.

As a school kid, Chris taught himself how to program on a home computer. Within two years, he was selling his own computer programs by mail order to customers throughout the UK.

Chris graduated with a BSc (Hons) degree in Computer Science from Queen's University, Belfast, and a PhD degree in Electronic Engineering from Dublin City University. After college, he was employed as a software consultant by Accenture and, later, as a senior researcher by Broadcom Éireann Research. Thereafter, he was appointed Vice President of Engineering at Massana, a leading-edge start-up company developing integrated circuits and software for data communications.

Today, Chris is an Associate Professor and Head of the School of Computer Science at University College Dublin (UCD), Ireland. He leads a research group focused on inventing novel algorithms for analysing real-world sensor data. His work has been published in leading peer-reviewed journals and presented at major international conferences.

Chris lives in Dublin with his wife and two children.

Contents

Introduction

'One for you. One for me. One for you. One for me.' You are in the school yard. The sun is shining. You are sharing a packet of sweets with your best friend. 'One for you. One for me.' What you didn't realize back then was that sharing your sweets in this way was an enactment of an algorithm.

An algorithm is a series of steps that can be performed to solve an information problem. On that sunny day, you used an algorithm to share your sweets fairly. The input to the algorithm was the number of sweets in the packet. The output was the number of sweets that you and your friend each received. If the total number of sweets in the packet happened to be even, then both of you received the same number of sweets. If the total was odd, your friend ended up with one sweet more than you.

An algorithm is like a recipe. It is a list of simple steps that, if followed, transforms a set of inputs into a desired output. The difference is that an algorithm processes information, whereas a recipe prepares food. Typically, an algorithm operates on physical quantities that represent information.

Often, there are alternative algorithms for solving a given problem. You could have shared your sweets by counting them, dividing the total by two in your head, and handing over the correct number of sweets. The outcome would have been the same, but the algorithm—the means of obtaining the output—would have been different.

An algorithm is written down as a list of instructions. Mostly, these instructions are carried out in sequence, one after another. Occasionally, the next instruction to be performed is not the next sequential step but an instruction elsewhere in the list. For example, a step may require the person performing the algorithm to go back to an earlier step and carry on from there. Skipping backwards like this allows repetition of groups of steps—a powerful feature in many algorithms. The steps, 'One for you. One for me.' were repeated in the sweet sharing algorithm. The act of repeating steps is known as *iteration*.

If the number of sweets in the packet was even, the following iterative algorithm would have sufficed:

Repeat the following steps:
  Give one sweet to your friend.
  Give one sweet to yourself.
Stop repeating when the packet is empty.

In the exposition of an algorithm such as this, steps are usually written down line-by-line for clarity. Indentation normally groups inter-related steps.

If the number of sweets in the packet could be even or odd, the algorithm becomes a little more complicated. A decision-making step must be included. Most algorithms contain decision-making steps. A decision-making step requires the operator performing the algorithm to choose between two possible courses of action. Which action is carried out depends on a *condition*. A condition is a statement that is either true or false. The most common decision-making construct— 'if-then-else'—combines a condition and two possible actions. 'If' the condition is true, 'then' the immediately following action (or actions) is performed. 'If' the condition is false, the step (or steps) after the 'else' are performed.

To allow for an odd number of sweets, the following decision-making steps must be incorporated in the algorithm:

If this is the first sweet or you just received a sweet,
then give this sweet to your friend,
else give this sweet to yourself.

The condition here is *compound*, meaning that it consists of two (or more) simple conditions. The simple conditions are 'this is the first sweet' together with 'you just received a sweet'. The two simple conditions are conjoined by an 'or' operation. The compound condition is true if either one of the simple conditions is true. In the case that the compound condition is true, the step 'give this sweet to your friend' is carried out. Otherwise, the step 'give this sweet to yourself' is performed.

The complete algorithm is then:

Take a packet of sweets as input.
Repeat the following steps:
    Take a sweet out of the packet.
    If this is the first sweet or you just received a sweet,
    then give this sweet to your friend,
    else give this sweet to yourself.
Stop repeating when the packet is empty.
Put the empty packet in the bin.
The sweets are now shared fairly.

Like all good algorithms, this one is neat and achieves its objective in an efficient manner.

The Trainee Librarian

Information problems crop up every day. Imagine a trainee librarian on their first day at work. One thousand brand new books have just been delivered and are lying in boxes on the floor. The boss wants the books to be put on the shelves in alphabetical order by author name, as soon as possible. This is an information problem and there are algorithms for solving it.

Most people would intuitively use an algorithm called Insertion Sort (Figure I.1). Insertion Sort operates in the following way:

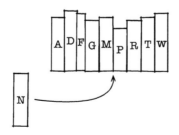

Figure I.1 Insertion Sort in action.

Take a pile of unsorted books as input.
Repeat the following steps:
    Pick up a book.
    Read the author's name.
    Scan across the shelf until you find where the book should
        be inserted.
    Shift all the books after that point over by one.
    Insert the new book.
Stop repeating when there are no books left on the floor.
The books are now sorted.

At any moment in time, the books on the floor are unsorted. One by one, the books are transferred to the shelf. Every book is placed on the shelf in alphabetical order. As a result, the books on the shelf are always in order.

Insertion Sort is easy to understand and works but is slow. It is slow because, for every book taken from the floor, the librarian has to scan past or shift every book already on the shelf. At the start, there are very few books on the shelf, so scanning and shifting is fast. At the end, our librarian has almost 1,000 books on the shelf. On average, putting a book in the right place requires 500 *operations*, where an operation is an author name comparison or a book shift. Thus, sorting all of the books takes 500,000 (1,000 × 500) operations, on average. Let's say that a single operation takes one second. That being the case, sorting the books using Insertion Sort will take around seventeen working days. The boss isn't going to be happy.

A faster alternative algorithm—Quicksort—was invented by computer scientist Tony Hoare in 1962. Hoare was born in Sri Lanka, to British parents in 1938. He was educated in England and attended Oxford University before entering academia as a lecturer. His method for sorting is a divide-and-conquer algorithm. It is more complicated than Insertion Sort but, as the name suggests, much faster.

Quicksort (Figure I.2) splits the pile of books into two. The split is governed by a *pivot letter*. Books with author names before the pivot letter are put on a new pile to the left of the current pile. Books with author names after the pivot are placed on a pile to the right. The resulting piles are then split using new pivot letters. In doing so, the piles are kept in sequence. The leftmost pile contains the books that come first in the alphabet. The next pile holds the books that come second, and so on. This pile-splitting process is repeated for the largest pile until

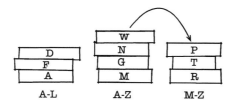

Figure I.2 Quicksort in action.

the biggest stack contains just five books. The piles are then sorted separately using Insertion Sort. Finally, the sorted piles are transferred, in order, to the shelf.

For maximum speed, the pivot letters should split the piles into two halves.

Let's say that the original pile contains books from A to Z. A good choice for the first pivot would likely be M. This would give two new piles: A–L and M–Z (Figure I.2). If the A–L pile is larger, it will be split next. A good pivot for A–L might be F. After this split, there will be three piles: A–E, F–L, and M–Z. Next, M–Z will be split and so on. For twenty books, the final piles might be: A–C, D–E, F–L, M–R, and S–Z. These piles are ordered separately using Insertion Sort and the books transferred pile-after-pile onto the shelf.

The complete Quicksort algorithm can be written down as follows:

Take a pile of unsorted books as input.
Repeat the following steps:
    Select the largest pile.
    Clear space for piles on either side.
    Choose a pivot letter.
    Repeat the following steps:
        Take a book from the selected pile.
        If the author name is before the pivot letter,
        then put the book on the pile to the left,
        else put the book on the pile to the right.
    Stop repeating when the selected pile is empty.
Stop repeating when the largest pile has five books or less.
Sort the piles separately using Insertion Sort.
Transfer the piles, in order, to the shelf.
The books are now sorted.

Quicksort uses two repeating sequences of steps, or *loops*, one inside the other. The outer repeating group deals with all of the piles. The inner group processes a single pile.

Quicksort is much faster than Insertion Sort for large numbers of books. The trick is that splitting a pile is fast. Each book need only be compared with the pivot letter. Nothing needs to be done to the other books—no author name comparisons, no book shifts. Applying Insertion Sort at the end of Quicksort is efficient since the piles are small. Quicksort only requires about 10,000 operations to sort 1,000 books. The exact number of operations depends on how accurately the pivots halve the piles. At one second per operation, the job takes less than three hours—a big improvement on seventeen working days. The boss will be pleased.

Clearly, an algorithm's speed is important. Algorithms are rated according to their *computational complexity*. Computational complexity relates the number of steps required for execution of an algorithm to the number of inputs. The computational complexity of Quicksort is significantly lower than that of Insertion Sort.

Quicksort is called a divide-and-conquer algorithm because it splits the original large problem into smaller problems, solves these smaller problems separately, and then assembles the partial solutions to form the complete solution. As we will see, divide-and-conquer is a powerful strategy in algorithm design.

Many algorithms have been invented for sorting, including Merge Sort, Heapsort, Introsort, Timsort, Cubesort, Shell Sort, Bubble Sort, Binary Tree Sort, Cycle Sort, Library Sort, Patience Sorting, Smoothsort, Strand Sort, Tournament Sort, Cocktail Sort, Comb Sort, Gnome Sort, UnShuffle Sort, Block Sort, and Odd-Even Sort. All of these algorithms sort data, but each is unique. Some are faster than others. Some need more storage space than others. A few require that the inputs are prepared in a special way. A handful have simply been superseded.

Nowadays, algorithms are inextricably linked with computers. By definition, a computer is a machine that performs algorithms.

The Algorithm Machine

As discussed, an algorithm is an abstract method for solving a problem. An algorithm can be performed by a human or a computer. Prior to execution on a computer, an algorithm must be encoded as a list of instructions that the computer can carry out. A list of computer instructions is called a *program*. The great advantage of a computer is that it can automatically execute large numbers of instructions one-after-another at high speed. Surprisingly, a computer need not support a great variety of instructions. A few basic instruction types will suffice. All that is needed are instructions for data storage and retrieval, arithmetic, logic, repetition, and decision-making. Algorithms can be broken down into simple instructions such as these and executed by a computer.

The list of instructions to be performed and the data to be operated on are referred to as the computer *software*. In a modern computer, software is encoded as electronic voltage levels on microscopic wires. The computer *hardware*—the physical machine—executes the program one instruction at a time. Program execution causes the input data to be processed and leads to creation of the output data.

There are two reasons for the phenomenal success of the computer. First, computers can perform algorithms much more quickly than humans. A computer can perform billions of operations per second, whereas a human might do ten. Second, computer hardware is *general-purpose*, meaning that it can execute any algorithm. Just change the software and a computer will perform a completely different task. This gives the machine great flexibility. A computer can perform a wide range of duties—everything from word processing to video games. The key to this flexibility is that the program dictates what the general-purpose hardware does. Without the software, the hardware is idle. It is the program that animates the hardware.

The algorithm is the abstract description of what the computer must do. Thus, in solving a problem, the algorithm is paramount. The algorithm is the blueprint for what must be done. The program is the precise, machine-executable formulation of the algorithm. To solve an information problem, a suitable algorithm must first be found. Only then can the program be typed into a computer.

The invention of the computer in the mid-twentieth century gave rise to an explosion in the number, variety, and complexity of algorithms. Problems that were once thought impossible to solve are now routinely dispatched by cheap computers. New programs are released on a daily basis, extending the range of tasks that computers can undertake.

Algorithms are embedded in the computers on our desktops, in our cars, in our television sets, in our washing machines, in our smartphones, on our wrists, and, soon, in our bodies. We engage a plethora of algorithms to communicate with our friends, to accelerate our work, to play games, and to find our soulmates. Algorithms have undoubtedly made our lives easier. They have also provided humankind with unprecedented access to information. From astronomy to particle physics, algorithms have enhanced our comprehension of the universe. Recently, a handful of cutting-edge algorithms have displayed superhuman intelligence.

All of these algorithms are ingenious and elegant creations of the human mind. This book tells the story of how algorithms emerged from the obscure writings of ancient scholars to become one of the driving forces of the modern computerized world.

1

Ancient Algorithms

Go up on to the wall of Uruk, Ur-shanabi, and walk around,
Inspect the foundation platform and scrutinise the brickwork!
Testify that its bricks are baked bricks,
And that the Seven Counsellors must have laid its foundations!
One square mile is city, one square mile is orchards,
one square mile is clay pits,
as well as the open ground of Ishtar's temple.
Three square miles and the open ground comprise Uruk.

Unknown author, translated by Stephanie Dalley
The Epic of Gilgamesh, circa 2,000 BCE [2]

The desert has all but reclaimed Uruk. Its great buildings are almost
entirely buried beneath accretions of sand, their timbers disintegrated.
Here and there, clay brickwork is exposed, stripped bare by the wind or
archaeologists. The abandoned ruins seem irrelevant, forgotten, futile.
There is no indication that seven thousand years ago, this land was the
most important place on Earth. Uruk, in the land of Sumer, was one of
the first cities. It was here, in Sumer, that civilization was born.

Sumer lies in southern Mesopotamia (Figure 1.1). The region is
bounded by the Tigris and Euphrates rivers, which flow from the
mountains of Turkey in the north to the Persian Gulf in the south.
Today, the region straddles the Iran–Iraq border. The climate is hot and
dry, and the land inhospitable, save for the regular flooding of the river
plains. Aided by irrigation, early agriculture blossomed in the 'land
between the rivers'. The resulting surplus of food allowed civilization
to take hold and flourish.

The kings of Sumer built great cities—Eridu, Uruk, Kish, and Ur.
At its apex, Uruk was home to sixty thousand people. All of life was
there—family and friends, trade and religion, politics and war. We know
this because writing was invented in Sumer around 5,000 years ago.

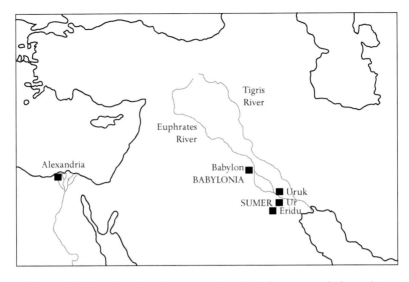

Figure 1.1 Map of ancient Mesopotamia and the later port of Alexandria.

Etched in Clay

It seems that writing developed from simple marks impressed on wet clay tokens. Originally, these tokens were used for record keeping and exchange. A token might equate to a quantity of gain or a headcount of livestock. In time, the Sumerians began to inscribe more complex patterns on larger pieces of clay. Over the course of centuries, simple pictograms evolved into a fully formed writing system. That system is now referred to as *cuneiform* script. The name derives from the script's distinctive 'wedge shaped' markings, formed by impressing a reed stylus into wet clay. Symbols consisted of geometric arrangements of wedges. These inscriptions were preserved by drying the wet tablets in the sun. Viewed today, the tablets are aesthetically pleasing—the wedges thin and elegant, the symbols regular, the text neatly organized into rows and columns.

The invention of writing must have transformed these communities. The tablets allowed communication over space and time. Letters could be sent. Deals could be recorded for future reference. Writing facilitated the smooth operation and expansion of civil society.

For a millennium, cuneiform script recorded the Sumerian language. In the twenty-fourth century BCE, Sumer was invaded by the armies of the Akkadian Empire. The conquerors adapted the Sumerian writing methods to the needs of their own language. For a period, both languages were used on tablets. Gradually, as political power shifted, Akkadian became the exclusive language of the tablets.

The Akkadian Empire survived for three centuries. Thereafter, the occupied city states splintered, later coalescing into Assyria in the north and Babylonia in south. In the eighteenth century BCE, Hammurabi, King of Babylon, reunited the cities of Mesopotamia. The city of Babylon became the undisputed centre of Mesopotamian culture. Under the King's direction, the city expanded to include impressive monuments and fine temples. Babylonia became a regional superpower. The Akkadian language, and its cuneiform script, became the *lingua franca* of international diplomacy throughout the Middle East.

After more than one millennium of dominance, Babylon fell, almost without resistance, to Cyrus the Great, King of Persia. With its capital in modern Iran, the Persian Empire engulfed the Middle East. Cyrus's Empire stretched from the Bosporus strait to central Pakistan and from the Black Sea to the Persian Gulf. Persian cuneiform script came to dominate administration. Similar at first glance to the Akkadian tablets, these new tablets used the Persian language and an entirely different set of symbols. Use of the older Akkadian script dwindled. Four centuries after the fall of Babylon, Akkadian fell into disuse. Soon, all understanding of the archaic Sumerian and Akkadian cuneiform symbols was lost.

The ancient cities of Mesopotamia were gradually abandoned. Beneath the ruins, thousands of tablets—records of a dead civilization lay buried. Two millennia passed.

Uncovered at Last

European archaeologists began to investigate the ruins of Mesopotamia in the nineteenth century. Their excavations probed the ancient sites. The artefacts they unearthed were shipped back to Europe for inspection. Amongst their haul lay collections of the inscribed clay tablets. The tablets bore writing of some sort, but the symbols were now incomprehensible.

Assyriologists took to the daunting task of deciphering the unknown inscriptions. Certain oft repeated symbols could be identified and decoded. The names of kings and provinces became clear. Otherwise, the texts remained impenetrable.

The turning point for translators was the discovery of the Behistun (Bīsitūn) Inscription. The Inscription consists of text accompanied by a relief depicting King Darius meting out punishment to handcuffed prisoners. Judging by their garb, these captives were from across the Persian Empire. The relief is carved high on a limestone cliff face overlooking an ancient roadway in the foothills of the Zagros mountains in western Iran. The Inscription is an impressive fifteen metres tall and twenty five metres wide.

The significance of the Inscription only became apparent after Sir Henry Rawlinson—a British East India Company officer—visited the site. Rawlinson scaled the cliff and made a copy of the cuneiform text. In doing so, he spotted two other inscriptions on the cliff. Unfortunately, these were inaccessible. Undaunted, Rawlinson returned in 1844 and, with the aid of a local lad, secured impressions of the other texts.

It transpired that the three texts were in different languages—Old Persian, Elamite, and Babylonian. Crucially, all three recounted the same propaganda—a history of the King's claims to power and his merciless treatment of rebels. Some understanding of Old Persian had persisted down through the centuries. Rawlinson compiled and published the first complete translation of the Old Persian text two years later.

Taking the Old Persian translation as a reference, Rawlinson and a loose cadre of enthusiasts succeeded in decoding the Babylonian text. The breakthrough was the key to unlocking the meaning of the Akkadian and Sumerian tablets.

The tablets in the museums of Baghdad, London, and Berlin were revisited. Symbol by symbol, tablet by tablet, the messages of the Sumerians, Akkadians, and Babylonians were decoded. A long-lost civilization was revealed.

The messages on the earliest tablets were simplistic. They recorded major events, such as the reign of a king or the date of an important battle. Over time, the topics became more complex. Legends were discovered, including the earliest written story: *The Epic of Gilgamesh*. The day-to-day administration of civil society was revealed—laws, legal contracts, accounts, and tax ledgers. Letters exchanged by kings and queens were found, detailing trade deals, proposals of royal marriage, and threats of war. Personal epistles were uncovered, including love poems and magical curses. Amid the flotsam and jetsam of daily life, scholars stumbled upon the algorithms of ancient Mesopotamia. Many of the extant Mesopotamian algorithms were jotted down by students learning mathematics. The following example dates from the Hammurabi dynasty (1,800 to 1,600 BCE), a time now known as the Old Babylonian period. Dates are approximate; they are inferred from the linguistic style of the text and the symbols employed. This algorithm was pieced together from fragments held in the British and Berlin State Museums. Parts of the original are still missing.

The tablet presents an algorithm for calculating the length and width of an underground water cistern. The presentation is formal and consistent with other Old Babylonian algorithms. The first three lines are a concise description of the problem to be solved. The remainder of the text is an exposition of the algorithm. A worked example is interwoven with the algorithmic steps to aid comprehension.[5]

A cistern.
The height is 3.33, and a volume of 27.78 has been excavated.
The length exceeds the width by 0.83.
You should take the reciprocal of the height, 3.33, obtaining 0.3.
Multiply this by the volume, 27.78, obtaining 8.33.
Take half of 0.83 and square it, obtaining 0.17.
Add 8.33 and you get 8.51.
The square root is 2.92.
Make two copies of this, adding to the one 0.42 and subtracting
    from the other.
You find that 3.33 is the length and 2.5 is the width.
This is the procedure.

The question posed is to calculate the length and width of a cistern, presumably of water. The volume of the cistern is stated, as is its height. The required difference between the cistern's length and width is specified. The actual length and width are to be determined.

The phrase, 'You should', indicates that what follows is the method for solving the problem. The result is followed by the declaration, 'This is the procedure', which signifies the end of the algorithm.

The Old Babylonian algorithm is far from simple. It divides the volume by the height to obtain the area of the base of the cistern.

Simply taking the square root of this area would give the length and width of a square base. An adjustment must be made to create the desired rectangular base. Since a square has minimum area for a given perimeter, the desired rectangle must have a slightly larger area than the square base. The additional area is calculated as the area of a square with sides equal to half the difference between the desired length and width. The algorithm adds this additional area to the area of the square base. The width of a square with this combined area is calculated. The desired rectangle is formed by stretching this larger square. The lengths of two opposite sides are increased by half of the desired length–width difference. The length of the other two sides is decreased by the same amount. This produces a rectangle with the correct dimensions.

Decimal numbers are used the description above. In the original, the Babylonians utilized *sexagesimal* numbers. A sexagesimal number system possesses sixty unique digits (0–59). In contrast, decimal uses just ten digits (0–9). In both systems, the weight of a digit is determined by its position relative to the fractional (or decimal) point. In decimal, moving right-to-left, each digit is worth ten times the preceding digit. Thus, we have the units, the tens, the hundreds, the thousands, and so on. For example, the decimal number 421 is equal to four hundreds plus two tens plus one unit. In sexagesimal, moving right-to-left from the fractional point, each digit is worth sixty times the preceding one. Conversely, moving left-to-right, each column is worth a sixtieth of the previous one. Thus, sexagesimal 1,3.20, means one sixty plus three units plus twenty sixtieths, equal to $63\frac{20}{60}$ or 63.333 decimal. Seemingly, the sole advantage of the Old Babylonian system is that thirds are much easier to represent than in decimal.

To the modern reader, the Babylonian number system seems bizarre. However, we use it every day for measuring time. There are sixty seconds in a minute and sixty minutes in an hour. The time 3:04 am is $184\,(3 \times 60 + 4 \times 1)$ minutes after midnight.

Babylonian mathematics contains three other oddities. First, the fractional point wasn't written down. Babylonian scholars had to infer its position based on context. This must have been problematic—consider a price tag with no distinction between dollars and cents! Second, the Babylonians did not have a symbol for zero. Today, we highlight the gap left for zero by drawing a ring around it (0). Third, division was performed by multiplying by the reciprocal of the divisor. In other words, the Babylonians didn't divide by two, they multiplied by a half.

In practice, students referred to precalculated tables of reciprocals and multiplications to speed up calculations.

A small round tablet shows the breathtaking extent of Babylonian mathematics. The tablet—YBC 7289—resides in Yale University's Old Babylonian Collection (Figure 1.2). Dating to around 1,800 to 1,600 BCE, it depicts a square with two diagonal lines connecting opposing corners. The length of the sides of the square are marked as thirty units. The length of the diagonal is recorded as thirty times the square root of two.

The values indicate knowledge of the Pythagorean Theorem, which you may recall from school. It states that, in triangles with a right angle, the square (a value multiplied by itself) of the length of the hypotenuse (the longest side) is equal to the sum of the squares of the lengths of the other two sides.

What is truly remarkable about the tablet is that it was inscribed 1,000 years before the ancient Greek mathematician Pythagoras was born. For mathematicians, the discovery is akin finding an electric light bulb in a Viking camp! It raises fundamental questions about the history of mathematics. Did Pythagoras invent the algorithm, or did he learn of it during his travels? Was the Theorem forgotten and independently reinvented by Pythagoras? What other algorithms did the Mesopotamians invent?

YBC 7289 states that the square root of two is 1.41421296 (in decimal). This is intriguing. We now know that the square root of two is 1.414213562, to nine decimal places. Remarkably, the value on the tablet

Figure 1.2 Yale Babylonian Collection tablet 7289. (*YBC 7289, Courtesy of the Yale Babylonian Collection.*)

is accurate to almost seven digits, or 0.0000006. How did the Babylonians calculate the square root of two so precisely?

Computing the square root of two is not trivial. The simplest method is Heron of Alexandria's approximation algorithm. There is, of course, the slight difficulty that Heron lived 1,500 years (c. 10–70 CE) after YBE 7289 was inscribed! We must assume that the Babylonians devised the same method.

Heron's algorithm reverses the question. Instead of asking 'What is the square root of two?', Heron enquires 'What number multiplied by itself gives two?' Heron's algorithm starts with a guess and successively improves it over a number of iterations:

Make a guess for the square root of two.
Repeatedly generate new guesses as follows:
   Divide two by the current guess.
   Add the current guess.
   Divide by two to obtain a new guess.
Stop repeating when the two most recent guesses are almost
   equal.
The latest guess is an approximation for the square root of two.

Let's say that the algorithm begins with the extremely poor guess of:

2.

Dividing 2 by 2 gives 1. Adding 2 to this, and dividing by 2 gives:

1.5.

Dividing 2 by 1.5 gives 1.333. Adding 1.5 to this and dividing by 2 again gives:

1.416666666.

Repeating once more gives:

1.41421568.

which is close to the true value.

How does the algorithm work? Imagine that you know the true value for the square root of two. If you divide two by this number, the result is exactly the same value—the square root of two.

Now, imagine that your guess is greater than the square root of two. When you divide two by this number, you obtain a value less than the square root of two. These two numbers frame the true square root— one is too large, the other is too small. An improved estimate can be obtained by calculating the average of these two numbers (i.e. the sum divided by two). This gives a value midway between the two framing numbers.

This procedure—division and averaging—can be repeated to further refine the estimate. Over successive iterations, the estimates converge on the true square root.

It is worth noting that the process also works if the guess is less than the true square root. In this case, the number obtained by means of division is too large. Again, the two values frame the true square root.

Even today, Heron's method is used to estimate square roots. An extended version of the algorithm was utilized by Greg Fee in 1996 to confirm enumeration of the square root of two to ten million digits.

Mesopotamian mathematicians went so far as to invoke the use of memory in their algorithms. Their command 'Keep this number in your head' is an antecedent of the data storage instructions available in a modern computer.

Curiously, Babylonian algorithms do not seem to have contained explicit decision-making ('if–then–else') steps. 'If–then' rules were, however, used by the Babylonians to systematize non-mathematical knowledge. The Code of Hammurabi, dating from 1754 BCE, set out 282 laws by which citizens should live. Every law included a crime and a punishment: [8]

> If a son strike a father, they shall cut off his fingers.
>
> If a man destroy the eye of another man, they shall destroy his eye.

If–then constructs were also used to capture medical knowledge and superstitions. The following omens come from the library of King Ashurbanipal in Nineveh around 650 BCE: [9]

> If a town is set on a hill, it will not be good for the dweller within that town.
>
> If a man unwittingly treads on a lizard and kills it, he will prevail over his adversary.

Despite the dearth of decision-making steps, the Mesopotamians solved a wide variety of problems by means of algorithms. They projected interest on loans, made astronomical predictions, and even solved quadratic equations (i.e. equations with unknowns to the power of two). While most of their algorithms had practical applications, a few suggest the pursuit of mathematics for its own sake.

Elegance and Beauty

The invention of writing in Egypt using hieroglyphs was roughly contemporaneous with its development in Mesopotamia. Due to the use of perishable papyrus scrolls, little evidence of Egyptian mathematics has survived to the present day. The most notable extant record is a papyrus scroll purchased by Henry Rhind in Luxor in 1858. Now in the British Museum, the Rind Papyrus is an ancient copy of an original dating from around 2,000 BCE. The five-metre-long and thirty-three-centimetre-wide roll poses a series of problems in arithmetic, algebra, and geometry. While the basics of these topics are well covered, little of the content is algorithmic in nature. Overall, it seems that algorithms were not a well-developed component of ancient Egyptian mathematics.

In the centuries after the rise of the Persian Empire, the Hellenic world gradually took up the mantle of leadership in mathematics. The Greeks learned much from the Mesopotamians and Egyptians as by-products of trade and war.

Alexander the Great (356–323 BCE) established Greek military dominance over the entire Middle East in the period 333–323 BCE. His conquests began with the unification of the Greek city states under his sole rule by means of military victory. Subsequently, the young man raised an army of 32,000 infantry allied with 5,000 cavalry and marched on Asia Minor. Alexander proved to be a brilliant military tactician and an inspirational leader. His force swept through Syria, Egypt, Phoenicia, Persia, and Afghanistan, taking city after city. Then, in 323 BCE, in the wake of one of his habitual drinking binges, the Emperor Alexander was taken with a fever. He died a few days later in Babylon, aged just 32. Alexander's vast empire was divided between four of his generals. Ptolemy—a close friend of Alexander's and, possibly, his half-brother—was appointed governor of Egypt.

One of Ptolemy's first decisions was to relocate Egypt's capital from Memphis to Alexandria. Alexander himself had founded the city on the site of an older Egyptian town. Alexandria was ideally located.

Situated on the Mediterranean coast at the western edge of the Nile delta, its natural harbours afforded the navy and commercial traders easy access to the Nile. Goods could be transported upriver by barge. Camel trains linked the Upper Nile with the Red Sea. Alexandria grew prosperous on trade. Buoyed by an influx of Egyptians, Greeks, and Jews, Alexandria become the largest city of its time. The historian Strabo described Alexandria thus: [11]

> The city also contains very beautiful public parks and royal palaces, which occupy a fourth or even a third of its whole extent.
>
> There follow along the waterfront a vast succession of docks, military and mercantile harbours, magazines, also canals reaching the lake Mareôtis, and many magnificent temples, an amphitheatre, stadium.
>
> In short, the city of Alexandria abounds with public and sacred buildings.

Ptolemy I commissioned construction of the Lighthouse of Alexandria. One of the seven wonders of the ancient world, the mighty lighthouse stood on the island of Pharos, which served as a bulwark between the open sea and the port. Elegant in design, the striking, three-tiered, 100-metre-tall, stone tower supported a beacon—mirror by day and fire by night—for shipping.

Ptolemy also founded a research institute known as the *Mouseion*, or Museum. The 'Seat of the Muses' was similar in nature to a modern research institute, attracting researchers, scientists, authors, and mathematicians from around the Mediterranean. Its most famous building was the renowned Library of Alexandria. The Library was intended to be a repository for all knowledge. Generously funded, it acquired one of the largest collections of scrolls in the world. Reputedly, at its zenith, the library held more than 200,000 books. It is said that all ships entering the harbour were searched for scrolls. Any material so uncovered was confiscated and a copy added to the Library. The Library of Alexandria became the preeminent centre of learning in the Mediterranean world.

Euclid (third–fourth century BCE) was perhaps the greatest Alexandrian scholar. Little is known of his life save that he opened a school in the city during the reign of Ptolemy I. Unfortunately, most of Euclid's writings are now lost. Copies of five of his books survive. His great work was *Euclid's Elements of Geometry*—a mathematics textbook. It drew on the writings of his predecessors, running to thirteen chapters covering geometry, proportions, and number theory. Down through the ages,

Euclid's Elements was copied, translated, recopied, and retranslated. What is now known as Euclid's algorithm is contained in Book VII.

Euclid's algorithm calculates the greatest common divisor of two numbers (also called the GCD, or the largest common factor). For example, 12 has six divisors (i.e. integers that divide evenly into it). The divisors are 12, 6, 4, 3, 2, and 1. The number 18 also happens to have six divisors: 18, 9, 6, 3, 2, and 1. The greatest common divisor of both 12 and 18 is therefore 6.

The GCD of two numbers can be found by listing all of their divisors and searching for the largest value common to both lists. This approach is fine for small numbers, but is very time consuming for large numbers. Euclid came up with a much faster method for finding the GCD of two numbers. The method has the advantage of only needing subtraction operations. Cumbersome divisions and multiplications are avoided.

Euclid's algorithm operates as follows:

Take a pair of numbers as input.
Repeat the following steps:
    Subtract the smaller from the larger.
    Replace the larger of the pair with the value obtained.
Stop repeating when the two numbers are equal.
The two numbers are equal to the GCD.

As an example, take the following two inputs:

12, 18.

The difference is 6. This replaces 18, the larger of the pair. The pair is now:

12, 6.

The difference is again 6. This replaces 12, giving the pair:

6, 6.

Since the numbers are equal, the GCD is 6.

It is not immediately obvious how the algorithm works. Imagine that you know the GCD at the outset. The two starting numbers must both be multiples of the GCD since the GCD is a divisor of both. Since both inputs are multiples of the GCD, the difference between them must also be a multiple of the GCD. By definition, the difference between the inputs has to be smaller than the larger of the two numbers. Replacing the larger number with the difference means that the pair of numbers is reduced. In other words, the pair is getting closer to the GCD. At all times, the pair, and their difference, are multiples of the GCD. Over several iterations, the difference becomes smaller and smaller. Eventually, the difference is zero. When this happens, the numbers are equal to the smallest possible multiple of the GCD, that is, the GCD times 1. At this point, the algorithm outputs the result and terminates.

This version of Euclid's algorithm is iterative. In other words, it contains repeating steps. Euclid's algorithm can, alternatively, be expressed *recursively*. Recursion occurs when an algorithm invokes itself. The idea is that every time the algorithm calls itself, the inputs are simplified. Over a number of calls, the inputs become simpler and simpler until, finally, the answer is obvious. Recursion is a powerful construct. The recursive version of Euclid's algorithm operates as follows:

Take a pair of numbers as input.
Subtract the smaller from the larger.
Replace the larger with the value obtained.
If the two numbers are equal,
then output one of the numbers – it is the GCD,
else apply this algorithm to the new pair of numbers.

This time, there is no explicit repetition of steps. The algorithm just calls for execution of itself. Each time, the algorithm is applied to a smaller pair of numbers: 18 and 12, then 12 and 6, next 6 and 6. Finally, the inputs are equal, and the result is returned.

The recursive version of Euclid's algorithm is one of the great algorithms. It is both effective and highly efficient. However, there is more to it than mere functionality. It has a symmetry, a beauty, and an elegance. Euclid's algorithm is an unexpected solution to the problem. It shows imagination and flair. All of these things make Euclid's algorithm great.

The great algorithms are poems that solve puzzles.

Finding Primes

In the third century BCE, Eratosthenes (c. 276–195 BCE) was appointed Director of the Library of Alexandria. Born in Cyrene, a North African city founded by the Greeks, Eratosthenes spent most of his early years in Athens. In middle age, he was called by Ptolemy III—grandson of Ptolemy I—to take charge of the great Library and tutor the King's son.

Today, Eratosthenes is most famous for having measured the circumference of the Earth. He discovered that at noon on the summer solstice (the longest day of the year), the shadow cast by a stake in the ground at Alexandria is longer than that cast by a stake of equal height at Syene (now Aswan), 800 km to the south. Measuring the distance between the cities gave Eratosthenes the length of the arc of the Earth between Alexandria and Syene. Combining this with the ratio of the shadow lengths, he produced an estimate the circumference of the Earth. Amazingly, his calculation of five times the distance between the two cities was accurate to within sixteen per cent of the true value.

As part of his researches on mathematics, Eratosthenes invented an important algorithm for finding prime numbers—the Sieve of Eratosthenes. A prime number has no exact whole number divisors (i.e. numbers that divide into it evenly) other than itself and one. The first five primes are 2, 3, 5, 7, and 11.

Primes are notoriously difficult to find. There are infinitely many, but they are scattered randomly across the number line. Even with modern computers, discovering new primes is time consuming. Some algorithms provide shortcuts but, to date, there is no easy way to find all primes.

The Sieve of Eratosthenes operates as follows:

List the numbers that you wish to find primes among, starting
    at 2.
Repeat the following steps:
    Find the first number that hasn't been circled or crossed out.
    Circle it.
    Cross out all multiples of this number.
Stop repeating when all the numbers are either circled or
    crossed out.
The circled numbers are prime.

Imagine trying to find all of the primes up to fifteen. The first step is to write down the numbers from 2 to 15. Next, circle 2 and cross out its multiples: 4, 6, 8, and so on.

(2), 3, 4, 5, 6, 7, 8, 9, 10, 11, 12, 13, 14, 15

Then, circle 3 and cross out all of its multiples: 6, 9, 12, 15.

②, ③, 4, 5, 6̶, 7, 8̶, 9̶, 1̶0̶, 11, 1̶2̶, 13, 1̶4̶, 1̶5̶

Four is already crossed out, so the next number to circle is 5, and so it goes. The final list is:

②, ③, 4, ⑤, 6̶, ⑦, 8̶, 9̶, 1̶0̶, ⑪, 1̶2̶, ⑬, 1̶4̶, 1̶5̶

The numbers that pass through the sieve (that is, are circled) are prime.

One of the neat aspects of the Sieve of Eratosthenes is that it does not use multiplication. Since the multiples are generated sequentially, one after another, they can be produced by repeatedly adding the circled number to a running total. For example, the multiples of 2 can be calculated by repeatedly adding two to a running total, giving 4, 6, 8, and so on.

A drawback of the Sieve is the amount of storage that it needs. To produce the first seven primes, eighteen numbers have to be stored. The amount of storage can be reduced by only recording whether a number has been struck off or not. Nevertheless, the *storage complexity* of the Sieve becomes a problem for large numbers of primes. An up-to-date laptop computer can find all primes with fewer than eight decimal digits using the Sieve of Eratosthenes. In contrast, as of March 2018, the largest known prime number contains a whopping 23,249,425 decimal digits.

For three hundred years, the Museum of Alexandria was a beacon of teaching and learning. Thereafter, slow decline was punctuated by disaster. In 48 BCE, Julius Caesar's army put their vessels to the torch in Alexandria's harbour in a desperate attempt to stall Ptolemy XIV's forces. The fire spread to the docks and parts of the Library were damaged in the resulting conflagration. The Museum was damaged in an Egyptian revolt in 272 CE. The Temple of Serapis was demolished in 391 CE by the order of Coptic Christian Pope Theophilus of Alexandria. The female mathematician Hypatia was murdered in 415 CE by a Christian mob. The Library was finally destroyed when General ʿAmr ibn al-ʿĀṣal–Sahmī's army took control of the city in 641 CE.

While the Museum of Alexandria was the pre-eminent centre of learning in the ancient Greek world for six centuries, it was not the only stronghold of logic and reason. On the other side of the Mediterranean, a lone genius invented a clever algorithm for calculating one of the most important numbers in all of mathematics. His algorithm was to outshine all others for nearly a thousand years.

2

Ever-Expanding Circles

How I wish I could recollect of circle round,
The exact relation Archimede unwound.
Unknown author, recounted by J. S. MacKay
Mnemonics for π, $\frac{1}{\pi}$, e, 1884 [13]

Göbekli Tepe lies in southern Turkey, close to the headwaters of the Euphrates river. Excavations at the site have uncovered a series of mysterious megalithic structures. Four-metre-high limestone pillars are set out in ten- to twenty-metre-wide circles. The circles are centred on pairs of still larger monoliths. The pillars are shaped similarly to an elongated 'T'. Most are richly engraved with depictions of animals. In places, the motifs are reminiscent of human hands and arms. In total, there are twenty circles and about 200 pillars.

While the structures are impressive, the truly extraordinary aspect of Göbekli Tepe is its age. The site dates from 10,000–8,000 BCE, far pre-dating ancient Sumer. This makes Göbekli Tepe the oldest known megalithic site in the world.

The practice of megalithic circle building was still in existence in Europe more than six thousand years later. One wonders what is so special about the circle that humankind chose to incorporate it in its greatest monuments over such a long period?

Wheels Within Wheels

The fundamental characteristic of the circle is that the distance from its centre to its perimeter is constant. This distance is the circle's *radius*. A circle's *diameter*, or width, is twice its radius. The *circumference* of a circle is the length of its perimeter. The larger a circle, the greater both its circumference and diameter. An assessment of the relationship between circumference and diameter can be made by measurement. Stretch a piece of string across a circle's diameter and compare that length to the circumference. You will find that the circle's circumference is slightly more than three times its diameter. Repeated measurement shows that this ratio is constant for all circle sizes. Of course, 'slightly more than three times' isn't particularly satisfactory from a mathematical point of view. Mathematicians want precise answers. Determining the precise ratio of a circle's circumference to its diameter is a never-ending quest.

The exact ratio—whatever its true value—is today represented by the Greek letter π (pronounced 'pi'). The letter π was first used in this way, not by the ancient Greeks, but by a Welshman—mathematician William Jones in 1707.

Writing down the true value of π is impossible. Johann Heinrich Lambert proved that π is an *irrational number*, meaning that enumeration requires infinitely many digits (1760s). No matter how far enumeration goes, the digits never settle into a repeating pattern. The best anyone can do is approximate π.

After the first few integers, π is arguably the most important number in mathematics. Without π, we would struggle to reason about circles and spheres. Circular motion, rotations, and vibrations would become mathematical conundrums. The value of π is employed in many practical applications ranging from construction to communication, and from spaceflight to quantum mechanics.

The original estimate of 3 is correct to 1 digit. By about 2,000 BCE, the Babylonians had estimated π as $\frac{25}{8} = 3.125$, accurate to two digits. The Egyptian Rhind Papyrus offers an improved approximation of $\frac{256}{81} = 3.16049$, close to three-digit accuracy. However, the first real breakthrough in determining π came from the Greek mathematician Archimedes.

Archimedes (c. 287–212 BCE) is considered to have been the greatest mathematician of antiquity. He was born in the city of Syracuse, Sicily, then part of a Greek colony.

The details of Archimedes' life are mostly unknown. Today, he is remembered for having leapt from his bathtub to run down the street yelling '*Eureka!*' ('I have found it!'). That tale comes from the histories of Vitruvius. It seems that Archimedes had been asked by the king to inspect the royal crown. The king was suspicious that his goldsmith had surreptitiously substituted a cheap silver–gold alloy for pure gold. The alloy looked identical to the real thing. Could Archimedes determine the truth?

There is one measurable difference between a silver–gold alloy and pure gold: pure gold is more dense. The density of an object is its weight (or mass) divided by its volume. The weight of the crown could be measured. However, determining its volume seemed impossible, as its shape was irregular.

One fateful night, Archimedes' bath happened to overflow as he climbed in. In a flash, Archimedes realised that the volume of an irregular object can be determined by measuring the quantity of water that it displaces when immersed. This deduction allowed him to determine the density of the crown.

The crown was not pure gold. The goldsmith was guilty.

Archimedes solved a series of important problems in mechanics, including defining the law of the lever. His greatest contributions, however, were in geometry. His studies led to him to enquire upon the correct value for π. The result of his labours was an algorithm for calculating π with unheard-of precision.

Archimedes' algorithm for approximating π is based on three insights. Firstly, a regular polygon approximates a circle. Secondly, it is easy to calculate the perimeter of a polygon because its sides are straight. Thirdly, the more sides a regular polygon has, the closer its approximation to a circle.

Imagine a circle. Now, visualize a hexagon (a regular six-sided figure) drawn just inside the circle (Figure 2.1). The corners of the hexagon touch the circle's perimeter and its sides lie just inside the circle. Since the hexagon is smaller than the circle, it stands to reason that the perimeter of the hexagon is close to, but slightly less than, that of the circle.

A regular hexagon is equivalent in outline to six identical triangles placed edge-to-edge pointing towards the centre. These triangles are *equilateral*, that is, all three sides are the same length. Since a hexagon has six edges, its perimeter is equal to six triangle sides. The diameter of a hexagon is equal to two triangle sides. Thus, the ratio of a hexagon's perimeter to its diameter is $\frac{6}{2} = 3$. Hence, 3 is a reasonable approximation for π.

Now, consider a hexagon drawn just outside the circle (Figure 2.1). In this case, the middle of each hexagon side touches the circle, not the corners. The diameter of the circle is now equal to twice the distance from the centre of the polygon to the middle of one side. The perimeter of this, larger, hexagon is $2\sqrt{3}$ times the diameter of the circle. This gives

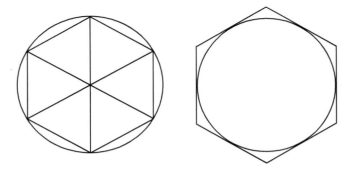

Figure 2.1 A circle with an inner hexagon (left) and a circle with an outer hexagon (right). The inner hexagon includes its constituent equilateral triangles.

another estimate for π equal to 3.46410. This estimate is close to the true value but is a little too large.

Archimedes improved these approximations by means of an algorithm. Every iteration of the algorithm doubles the number of sides in the two polygons. The more sides a polygon has, the better its approximation to π.

The algorithm operates as follows:

Take the perimeters of a pair of inner and outer polygons as input.
Multiply the inner and outer perimeters.
Divide by their sum.
This gives the perimeter of a new outer polygon.
Multiply this new outer perimeter by the previous inner perimeter.
Take the square root.
This gives the perimeter of a new inner polygon.
Output the perimeters of the new inner and outer polygons.

In the first iteration, the algorithm turns the hexagons into dodecagons (12-sided figures). This gives improved estimates for π of 3.10582 (inner polygonal) and 3.21539 (outer polygon) to six digits.

The beauty of Archimedes' algorithm is that it can be applied again. The outputs from one run can be fed into the algorithm as inputs to the next iteration. In this way, the dodecagons can be transformed into twenty-four-sided polygons. Forty-eight-sided polygons can be turned into ninety-six-sides figures, and so on. With every repetition, the inner and outer polygons close in on the circle, providing better estimates for π.

Archimedes completed the calculations for a ninety-six-sided figure, obtaining estimates for π of $\frac{223}{71}$ and $\frac{22}{7}$. The former is accurate to four digits. The latter is less accurate but more popular due to its simplicity.

Tragically, Archimedes was slain by a Roman soldier during the sack of Syracuse. Accounts differ as to the provocation. In one telling, Archimedes declined to accompany the soldier to his superior officer on the grounds that he was working on a particularly intriguing problem. In another, Archimedes attempted to prevent the soldier stealing his scientific instruments. Amazingly, it would be nearly two thousand years before Archimedes' algorithm was surpassed (1699 CE).

World Records

Archaeological evidence suggests that civilization emerged in China at around the same time as it appeared in Mesopotamia and Egypt. Urban society in China appears to have first developed along the banks of the Yangtze and Yellow Rivers. Little is known of early Chinese mathematics as the bamboo strips used for writing at the time were perishable. Although there was intercommunication between East and West, it appears that Chinese mathematics developed largely independently.

The oldest extant Chinese mathematical text—*Zhoubi Suanjing*—dates to around 300 BCE. The book focuses on the calendar and geometry, and includes the Pythagorean Theorem. A compendium of mathematical problems analogous to the Rhind Papyrus—*Nine Chapters on the Mathematical Art*—survives from roughly the same period.

It seems that the search for π was much more determined in China than in the West. In 264 CE, Liu Hui used a ninety-six-sided inner polygon to obtain an approximation of 3.14—accurate to three figures. He later extended his method to a polygon of 3,072 sides, obtaining an improved estimate of 3.14159—accurate to six figures.

Zu Chongzhi (430–501 CE), assisted by his son Zu Chengzhi, produced an even better estimate in the fifth century AD. The father and son duo used a polygonal method similar to Archimedes' approach. However, they persevered for many more iterations. Their upper and lower bounds of 3.1415927 and 3.1415926 were accurate to seven digits—a new world record. Their accomplishment was to stand for nigh on nine hundred years, a testament to their dedication

Today, we know that π is equal to 3.14159265359 to twelve figures. Enumeration of π is now an endeavour for computer algorithms. According to the Guinness Book of World Records, the most accurate value of π is thirty-one trillion digits long. The programme that produced the value was written by Emma Haruka Iwao of Japan. Running on twenty-five virtual machines in the Google Cloud, the calculation took 121 days.

The Art of Reckoning

Archimedes' murder in 212 BCE was a harbinger of the Roman domination of Europe. Ancient Greece fell to the might of Rome in 146 BCE. From the first century BCE to the 5th century CE, the Roman Empire controlled the Mediterranean. When the Empire finally fell, European civilization disintegrated. The flame of European mathematics flickered and guttered for a thousand years. Amidst the darkness, a few centres of learning illuminated the East.

Caliph Harun al-Rashid founded the House of Wisdom (*Bayt al-Hikmah*) in his new capital of Baghdad around 762 CE. Expanded by his successors, the House grew to become a major intellectual centre from the ninth to the thirteenth century, a period now known as the Islamic Golden Age. Scholars working in the House translated scientific and philosophical texts written in Greek and Indian into Arabic. They also conducted original research in mathematics, geography, astronomy, and physics.

The House of Wisdom's most influential intellectual was Muhammad ibn Mūsā al-Khwārizmī. Al-Khwārizmī lived in Baghdad from around 780 to 850. Not much is known of his life. However, copies of three of his major works survive.

The Compendious Book on Calculation by Completion and Balancing focuses on algebra. In fact, we derive the English word 'algebra' from the book's Arabic title (*al-jabr*, meaning completion). The book describes how to use algorithms to solve mathematical problems, especially linear and quadratic equations. While algebra had been described previously, it was al-Khwārizmī's style of presentation that caught the eye. His treatment was more systematic, more step-by-step, more algorithmic, than that found in other works.

al-Khwārizmī's text *On the Hindu Art of Reckoning* (*c.* 825) describes the decimal number system, including the numerals that we employ today. The system's roots lie in the Indus Valley civilization (now southern Pakistan), which blossomed around 2,600 BCE—roughly contemporaneous with the construction of the pyramids in Giza. Little is known of the original mathematics of the region, save what can be gleaned from religious texts. Inscriptions suggest that the nine non-zero Hindu–Arabic numerals (1–9) appeared in the region between the third century BCE and the second century CE. Certainly, there is a clear reference to the nine non-zero Hindu numerals in a letter from Bishop Severus Sebokht, who lived in Mesopotamia around 650 CE. The numeral for zero (0) finally appeared in India at around the same time.

By the eighth century, many Persian scholars had adopted the Hindu–Arabic number system by reason of its great convenience— hence, al-Khwārizmī's book on the subject. His text became a conduit for the Hindu–Arabic numerals in their transfer to the West. *On the Hindu Art of Reckoning* was translated in 1126 from Arabic into Latin by Adelard of Bath—an English natural philosopher. Adelard's translation was followed by Leonardo of Pisa's (Fibonacci) book on the topic—*Liber Abaci*—in 1202.

In 1258, four hundred years after al-Khwārizmī's death, The House of Wisdom was destroyed in the Mongol sack of Baghdad.

Surprisingly, uptake of the new number system was slow. It would be centuries before Roman numerals (I, II, III, IV, V, ...) were displaced by the Hindu–Arabic digits. It seems that European scholars were perfectly happy to perform calculations on an abacus and record the results using Roman numerals. Decimal numbers only became the preferred option in the sixteenth century with the transition to pen and paper calculation.

It is al-Khwārizmī's name in the title of a Latin translation of his book—*Algoritmi de Numero Indorum*—that gives us the English word 'algorithm'.

Waves Upon Waves

The European Renaissance of the fourteenth to seventeenth centuries saw the rediscovery of classical philosophy, literature, and art. Mathematics, too, was resurgent, particularly in its application to practicalities such as accountancy, mechanics, and map-making. The invention of the printing press in the fifteenth century further spurred the spread of scholarship and learning.

The ensuing Enlightenment of the eighteenth century saw a revolution in Western philosophy. Centuries of dogma were swept away by the strictures of evidence and reason. Mathematics and science became the foundations of thought. Technological progress altered the very fabric of society. Democracy and the pursuit of individual liberty were in the ascendency.

Changing attitudes, heavy taxation, and failed harvests were to ignite the French Revolution in 1789. Amidst the bloody upheaval, a French mathematician laid the theoretical foundations for what was to become one of the world's most frequently used algorithms.

In 1768, Jean-Baptiste Joseph Fourier (Figure 2.2) was born in Auxerre, France. Orphaned at age nine, Fourier was educated in local schools run by religious orders. The lad's talent for mathematics became obvious as he entered his teenage years. Nonetheless, the young man undertook to train for the priesthood. On reaching adulthood, Fourier abandoned the ministry to devote his career to mathematics, taking up employment as a teacher. Soon, he became entangled in the political upheaval sweeping the country. Inspired by the ideals of the

Figure 2.2 Bust of Joseph Fourier by Pierre-Alphonse Fessard, 1839. (© *Guillaume Piolle* / *CC-BY 3.0.*)

Revolution, Fourier turned to political activism and joined the Auxerre Revolutionary Committee. In the Terror that followed, Fourier found himself implicated in a violent dispute between rival factions. He was imprisoned and only narrowly escaped the guillotine.

Subsequently, Fourier moved to Paris to study as a teacher. His mathematical aptitude led him to be appointed as a member of faculty at the newly established École Polytechnique. A mere two years later, he was Chair of Analysis and Mechanics. Fourier seemed destined for a life in academia until an unexpected turn of events changed the course of his life.

Fourier was appointed scientific advisor to Napoléon's army for the invasion of Egypt. The French forces took Alexandria on 1 July 1798. Thereafter, victory turned to ignominious defeat. Napoléon departed Egypt, but Fourier stayed on in Cairo. There, he dispensed his scientific duties and, in his spare time, conducted research on the antiquities of Egypt.

On Fourier's eventual repatriation, Napoléon arranged for the mathematician to be appointed Prefect (administrator) of the Department of Isère—the Alpine region centred on the city of Grenoble. It was here that Fourier embarked on his magnum opus. His *Théorie Analytique de la Chaleur* was published in 1822. Nominally it was about the conduction of heat in metal bars. More importantly, the book suggested that any waveform can be viewed as the summation of appropriately delayed and scaled harmonic waves. The idea was highly controversial at the time. Nevertheless, Fourier's postulate was subsequently proven to be correct.

To understand Fourier's proposal, we will conduct a thought experiment. Imagine a swimming pool. Put a wave machine at one end and assume that the opposite end has some sort of overflow that doesn't reflect waves. Let us say that the machine produces a single wave that is the length of the pool. We watch the crest move from one end of the pool to the other before the next crest appears back at the start. This simple waveform is called the *first harmonic* (Figure 2.3). Its *period*—the distance between two crests—is equal to the length of the pool.

Now, image that the wave machine moves faster. This time, two complete cycles of the waveform fit into the length of the pool. In other words, we can always see two crests, not just one. This waveform is the *second harmonic*. Its period is equal to half the length of the pool.

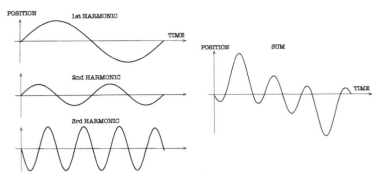

Figure 2.3 Three harmonics (left) and the waveform resulting from their summation (right). The second harmonic is scaled by a half and the third harmonic is delayed by a quarter cycle.

Next, we double the speed of the machine again. This time, the period is a quarter of the length of the pool. This is the *third harmonic*.

Another doubling and we get the *fourth harmonic*, and so on.

This sequence of harmonics is called the *Fourier series*.

Fourier's remarkable idea was that all waveforms—of any shape whatsoever—are the sums of scaled and delayed harmonics. Scaling means increasing or decreasing the size of the waveform. Scaling up makes the peaks higher and the troughs lower. Scaling down does the converse. Delaying a waveform means shifting it in time. A delay means that the wave's crests and troughs arrive later than before.

Let us examine the effect of combining harmonics (Figure 2.3). Let us say that the first harmonic has an *amplitude* of one. The amplitude of a waveform is its maximum deviation from rest. The amplitude of a harmonic is the height of the crests. The second harmonic has an amplitude of a half. The third harmonic has an amplitude of one and a delay of half a period. If we add these harmonics up, we get a new *compound waveform*. The process of addition mimics what happens in the real world when waves meet. They simply ride on top of one another. The terminology used in physics is to say that the waves *superpose*.

Clearly, adding waveforms is easy. Reversing the process is much more complicated. How, given a compound waveform, can the amplitudes and delays of the constituent harmonics be determined? The answer is by means of the *Fourier transform* (FT).

The FT takes any waveform as input and breaks it up into its component harmonics. Its output is the amplitude and delay of every harmonic in the original waveform. For example, given the compound waveform in Figure 2.3, the FT will output the amplitudes and delays of the three constituent harmonics. The output of the FT is two sequences, one of which lists the amplitudes of the harmonics. For the compound waveform in Figure 2.3, the amplitudes of the constituent harmonics are $[1, \frac{1}{2}, 1]$. The first entry is the amplitude of the first harmonic, and so on. The second output sequence is the delays of the harmonics: $[0, 0, \frac{1}{2}]$ measured in periods.

While initially only of interest to physicists, the real power of the FT became evident in the decades after the invention of the computer. Computers allow all kinds of waveforms to be quickly and cheaply analysed.

A computer stores a waveform as a list of numbers (Figure 2.4). Every number indicates the level of the waveform at a particular moment in time. Large positive values are associated with the peaks of the waveform. Large negative numbers with the troughs. These numbers are called *samples* since the computer takes a 'sample' of the level the waveform at regular intervals of time. If the samples are taken often enough, the list of numbers gives a reasonable approximation to the shape of the waveform.

The FT commences by generating the first harmonic. The generated waveform contains a single period—one crest and one trough—and is the same length as the input sequence. The algorithm multiplies the two lists of numbers sample-by-sample (e.g. $[1, -2, 3] \times [10, 11, 12] = [10, -22, 36]$). These results are then totalled. The total is the *correlation* of the input and the first harmonic. The correlation is a measure of the similarity of two waveforms. A high correlation indicates that the first harmonic is strong in the input.

The algorithm repeats the correlation procedure for a copy of the first harmonic delayed by a quarter of a period. This time, the corre-

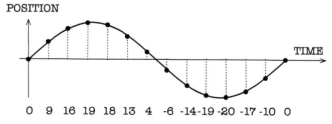

Figure 2.4 A waveform and the associated sample values that are used to represent the signal.

lation measures the similarity between the input waveform and the delayed first harmonic.

The two correlation values (not delayed and delayed) are fused to produce an estimate of the amplitude and delay of the first harmonic. The amplitude is equal to the square root of the sum of the squares of the two correlations divided by the number of samples. The delay is obtained by calculating the relative strength of the two correlations. The relative strength indicates how close the component is in time to the two versions of the harmonic.

This double correlation (not delayed and delayed) and fusion process is repeated for all higher harmonics. This gives their amplitudes and delays.

In summary, the FT algorithm works as follows:

Take a waveform as input.
Repeat the following steps for every possible harmonic:
    Generate the harmonic.
    Correlate the input and harmonic.
    Delay the harmonic by a quarter period.
    Correlate the input and delayed harmonic.
    Calculate the overall amplitude and delay of the harmonic.
Stop repeating when all harmonics have been processed.
Output the amplitude and delay of every harmonic.

On first inspection, this process of breaking a compound waveform into its constituent harmonics sounds like a mathematical conceit—a nice thing to demonstrate but of little practical use. This view could not be further from the truth! The FT is used extensively in the analysis of real-world *signals*.

A signal is any real-world quantity that varies with time. Sound signals are variations in air pressure that we can hear. In speech recognition systems (e.g. Siri, Alexa), the FT breaks the sound signal up into its constituent harmonics prior to further analysis. Digital music players (e.g. Spotify, Tidal), rely on the FT to identify redundant harmonic information and so reduce data storage requirements.

Radio signals are electromagnetic variations that can be detected using electronic equipment. In wireless communication systems (e.g. WiFi, DAB), the FT allows data to be efficiently transmitted and received by means of radio signals.

While the FT is extremely useful, it requires a lot of computing power. The correlations take an inordinate amount of time, particularly for long waveforms. A variant of the original algorithm—appropriately named the fast Fourier transform, or FFT—is used in modern devices.

The FFT was invented in 1965 to meet a pressing military need. With the Cold War on the horizon, the US wished to monitor Soviet nuclear tests. The only way to do so was to measure blast induced seismic vibrations at monitoring stations sited in friendly countries. Unfortunately, analysis of the seismic data using the traditional FT was prohibitively slow. Scientists James Cooley (1926–2016) and John Tukey (1915–2000) conjured up a new version of the algorithm that produced the same results in a fraction of the time.

Their algorithm exploits the symmetries in harmonic waveforms to share results between correlation stages. In a harmonic waveform, a trough is simply a reflection of a peak. The rising half of a crest is a mirror image of the falling part. Higher harmonics are accelerated version of lower harmonics. By re-using intermediate results, unnecessary repetitions in the calculation can be eliminated.

The idea worked a treat. The Cooley–Tukey FFT allowed the US military to locate Soviet nuclear tests to within fifteen kilometres.

Unexpectedly, almost twenty years after the 'invention' of the FFT, it was revealed that the algorithm was, in fact, more than 180 years old. The great German mathematician Carl Friedrich Gauss (1777–1855) had employed the algorithm in the analysis of astronomical data in 1805. Ever the perfectionist, Gauss never got around to publishing it. The method was finally uncovered amidst a posthumous collection of Gauss's papers. Gauss's 1805 notes even pre-date Fourier's work on the topic. In retrospect, it seems that the algorithm might more properly have been called the Gauss transform!

Fourier died on 16 May 1830. His name was later engraved on the side of the Eiffel Tower in recognition of his scientific achievements.

Fourier's turbulent life coincided almost exactly with the years of the Industrial Revolution. Over the course of a mere seventy years, ancient handcrafts were supplanted by machine production. Amidst the changes, an Englishman got to wondering whether a machine could weave calculations, rather than cloth. Might there be an Industrial Revolution for arithmetic?

3

Computer Dreams

The most obvious and the most distinctive features of the History
of Civilization, during the last fifty years, is the wonderful increase
of industrial production by the application of machinery, the
improvement of old technical processes and the invention of
new ones.

THOMAS HENRY HUXLEY
The Advance of Science in the Last Half-Century, 1887[25]

Performing calculations manually is slow and tedious. For millennia,
inventors have sought to design devices that accelerate arithmetic. The
first success was the abacus. Invented by the Sumerians around 2,500 BCE,
the table abacus evolved from counting with pebbles and sand writing.
Later, lines and symbols were etched into the table surface to facilitate
more rapid computation. The bead-and-rail abacus was invented in
China before being popularized in Europe by the ancient Romans.

The first mechanical calculators were invented independently in
France and Germany in the seventeenth century. Driven by hand
cranks, additions and subtractions were performed by the movement
of levers, cogs, and gears. The intricate hand-crafted devices were both
expensive and unreliable. Most were sold to the wealthy as mere curios.

The end of the eighteenth and the beginning of the nineteenth
century brought the Industrial Revolution. Engineers devised machines
powered by steam and running water that replaced traditional manual
production methods. The transition to machine production led to rapid
increases in productivity and major societal changes. Labourers were
displaced from the countryside to the towns and cities to work alongside
the machines in noisy, dark, and often dangerous factories.

Powered textile looms produced fabric with a fixed weave. In 1804,
Joseph Marie Charles ('Jacquard')—a French weaver and merchant—
came up with a radical redesign. Jacquard's loom could be *programmed*

to produce cloth with different weaves. The machine wove fabric according to the pattern of holes cut into cards. By altering the card punch pattern, a different weave could be produced. Sequences of cards were linked together in loops such that the machine could repeat the programmed pattern.

Thus, by the early nineteenth century, Europe was in possession of hand-powered calculators and steam-powered programmable looms. An English mathematician with a childhood fascination for machines began to wonder about combining these concepts. Surely, a steam-powered programmable machine could perform calculations much faster, and more reliably, than any human? His idea almost changed the world.

A Clockwork Computer

Charles Babbage was born in England in Walworth, Surrey in 1791 (Figure 3.1). The son of a wealthy banker, Babbage became a student

Figure 3.1 Charles Babbage (left, c. 1871) and Ada Lovelace (right, 1836), the first programmers. (*Left: Retrieved from the Library of Congress, www.loc.gov/item/2003680395/. Right: Government Art Collection. GAC 2172, Margaret Sarah Carpenter: (Augusta) Ada King, Countess of Lovelace (1815–1852) Mathematician; Daughter of Lord Byron.*)

of mathematics. Self-taught initially, he was admitted to Cambridge University at eighteen. Once there, he found the Mathematics department to be rather staid and disappointing. A headstrong young man, Babbage did not bother to ingratiate himself with either his examiners or his prospective employers. Despite being a superb mathematician, he failed to find a position in academia upon graduation. Supported by an allowance from his father, Babbage resolved to conduct his own independent mathematical research. He moved to London and entered into the scientific life of the capital, publishing a series of well-regarded papers.

Scientific papers, like those written by Babbage, are the lifeblood of research. A paper is a report describing a new idea backed up with supporting experimental results or a mathematical proof, ideally both. Science relies on evidence. Proofs must be verified. Experiments must be repeatable. Papers are reviewed by experts in the field. Only the best are accepted for publication. Crucially, the idea presented in a paper must be novel and proven. Publication in journals, or at conferences, propagates new ideas to interested parties in the scientific community. Publication of a paper is a milestone in an academic's career—an indication of their prowess and standing in the field.

Lacking public funding, science was conducted in a handful of universities and by a small number of wealthy enthusiasts. Scientific discourse often took place in the salons of the rich. Even the word 'scientist' was new. For years, Babbage was the quintessential Victorian gentleman scientist. He was eventually appointed Lucasian Professor of Mathematics at Cambridge in 1828. His innate talents were greatly amplified by his capacity for long, arduous working hours. Although an accomplished mathematician, perhaps his primary gift lay in invention of mechanical devices.

By dint of his published contributions, Babbage was elected a Fellow of the Royal Society. One of the duties that befell him was reviewing mathematical tables for the Astronomical Society. The tables listed the predicted times and positions of notable celestial events. These tables were used extensively by seafarers as an aid to navigation. Laborious to produce by means of manual calculation, the tables often contained errors. On the high seas, an error could lead to shipwreck.

To alleviate the workload in their production, Babbage produced a design for a steam-powered mechanical machine that could automatically calculate and print the tables. Decimal numbers were to be represented by the positions of gears, cogs, and levers. The engine would be capable of automatically performing a sequence of calculations, storing, and re-using intermediate results along the way. The machine was designed to perform a single, fixed algorithm. Hence, it lacked programmability. Nevertheless, the design was a significant advance on previous calculators, which required manual entry of each and every number and operation. Babbage fabricated a small working model of the machine. Seeing merit in Babbage's concept, the British government agreed to fund construction of Babbage's Difference Engine.

Building the Engine proved challenging. Tiny inaccuracies in fabrication of its components made the machine unreliable. Despite repeated investment by the government, Babbage and his assistant Jack Clement only completed a portion of the machine before construction was abandoned. In total, the British Treasury spent nearly £17,500 on the project. No small sum, the amount was sufficient to otherwise procure twenty-two brand new railway locomotives from Mr. Robert Stephenson.[26]

Despite the failure of his Difference Engine project, Babbage was still drawn to the idea of automated calculation. He designed a new, much more advanced machine. The Analytic Engine was to be mechanical, steam-driven, and decimal. It would also be fully programmable. Borrowing from Jacquard's loom, the new machine would read instructions and data from punch cards. Likewise, results would be proffered on punched cards. The Analytic Engine was to be the first general-purpose computer.

Once again, Babbage appealed to the government for funding. This time, the money was not forthcoming. The Analytic Engine project stalled.

Babbage made his only public presentation on the Analytic Engine to a group of mathematicians and engineers in Torino, Italy. One of the attendees—Luigi Federico Menabrea—a military engineer, made notes and, subsequently, with the help of Babbage, published a paper on the device. That paper was in French. Another supporter of Babbage's—Ada Lovelace—greatly admired the work and resolved to translate it into English.

Ada Lovelace (born Augusta Ada Byron) was born in 1815, the daughter of Lord and Lady Byron (Figure 3.1). Lady Byron (Anne

Isabella Noel, née Milbanke) was herself a mathematician. Her husband, Lord Byron, is still regarded as one of the great English poets. Lady and Lord Byron's marriage only lasted a year before the pair separated. Lady Byron told of her husband's dark moods and her mistreatment.[26] Rumours persisted of his infidelity. Disgraced, the poet left England, never to see his daughter again.

At her mother's bidding, Lovelace grew up studying science and mathematics. These were sensible subjects, far from the worrisome influence of poetry and literature. Aged only seventeen, she met Babbage in 1833 at one of his social gatherings. At the time, Babbage was a forty-one-year-old widower with four surviving children. Having inherited his father's fortune, he was living in style in a London mansion at 1 Dorset Street, Marylebone. Babbage's soirées were notable occasions—a heady mix of high society, artists, and scientists. Gatherings of two hundred or more notables were common. One wag had it that wealth alone was insufficient to procure an invitation. One of three qualifications was required: 'intellect, beauty, or rank'.

Lovelace was fascinated by Babbage's computing devices. At his invitation, she and her mother inspected the working portion of the Difference Engine. Babbage and Ada struck up a friendship. They corresponded regularly, discussing the Analytic Engine and other scientific topics. At nineteen, Ada married William King to become Augusta Ada King, Countess of Lovelace.

Not only did Ada Lovelace translate Menabrea's paper into English, she extended it with seven notes, more than doubling its length. The paper—*Sketch of the Analytical Engine*—was visionary. Although the Analytic Engine was never built, Babbage had specified the instructions that the envisaged machine would perform. This allowed Babbage and Lovelace to write programs for the non-existent computer.

In the paper, the authors emphasized the relationship between an algorithm and its equivalent program. They explained that an algorithm is an abstract calculation method written down, in Babbage and Lovelace's case, as a sequence of mathematical equations. The paper drew out the equivalence between the algorithm expressed as equations—or algebraic formulae, as they put in—and the program encoded on punched cards:[28]

> The cards are merely a translation of algebraic formulae, or, to express it better, another form of analytical notation.

The sequence of mathematical equations could only be performed by a human. In contrast, the instructions on the punched cards could be carried out automatically at high speed by the Analytic Engine.

Lovelace's addendum to the paper presented a series of numerical algorithms encoded as programs. Lacking an Analytic Engine, the only way to test a program was to execute it manually, mimicking the actions of the computer. In Note G of the paper, Lovelace provides an execution *trace* of a program for calculating the first five Bernoulli numbers. The trace lists the instructions as they would have been executed by the computer, together with the results obtained after every step. Unbeknownst to Lovelace, her trace echoed the Old Babylonian presentation of algorithms in tandem with worked examples. Modern-day programmers still use traces to better understand the behaviour of their programs.

Prophetically, the authors noted the prospect of mistakes, or *bugs*, in programs: [28]

> Granted that the actual mechanism is unerring in its processes, the cards may give it wrong orders.

Ironically, an error has since been found in one of the program listings included in the paper: a three where there should have been a four. This was the first bug in a software release. The first of a great many!

The paper also put forth the idea that the computer might process other forms of information, not just numbers: [28]

> It might act upon other things besides numbers, were objects found whose mutual fundamental relations could be expressed by those of the abstract science of operations, and which should be also susceptible of adaptations to the action of the operating notation and mechanism of the engine.

In other words, while the Engine was designed to perform arithmetic, the symbols employed could represent other forms of information. Furthermore, the machine could be programmed, or modified, so as to process these other forms of data. The authors understood that the machine could manipulate symbols representing letters, words, and logical values. Babbage even toyed with writing programs to play Tic-Tac-Toe and Chess.

Babbage, Menabrea, and Lovelace's spectacular paper was the first glimmer of a new science. A science at the nexus of algorithms, programming, and data: the science of computers. It would be another 100 years before the field was recognized and named as computer science.

This was to be Lovelace's only scientific publication. She suffered from poor health for many years, and aged only thirty-six, she died from cancer. She and Babbage remained close friends until the end. At her request, she was buried alongside her estranged father in Hucknall, Nottinghamshire.

Babbage worked intermittently on the Analytic Engine for the rest of his days. Time and again, his plans were frustrated by the deficiencies of the mechanical fabrication technologies of the day. Babbage's Analytic Engine was another glorious failure.

Nonetheless, Babbage was tireless. He travelled widely. On one occasion, he descended into Mount Vesuvius with the goal of surveying its volcano. He authored scholarly works on economics, geology, life assurance, and philosophy. Even aside from his computing engines, he was a prolific inventor. He dabbled in politics to the extent of standing for election. In his later years, Babbage became embittered by his seeming lack of recognition by the establishment. He engaged in a public campaign of words against the street musicians of London whose cacophony he found insufferable. Babbage died in 1871, eighteen years after Lovelace, and his dream of a mechanical digital computer died with him.

With hindsight, it is clear that the Analytic Engine had all of the ingredients of the modern computer, except one. It was mechanical, not electronic. The commercially manufactured electric light bulb was invented by Thomas Edison after Babbage's death. It would be another fifty years before anyone attempted to build a programmable electronic computer. Babbage's legacy to computing was Lovelace's paper.

Without a programmable Engine, it was up to the theoreticians to chart the future of algorithms and computation. One of the most influential of these was Alan Turing (Figure 3.2).

The Turing Machine

Turing was born in 1912 in London, England. As a consequence of his father being a colonial civil servant, Turing's parents returned to India when Alan was just one year old. Turing and his brother remained

Figure 3.2 Computing pioneer Alan Turing, c. 1930s.

in England in the care of a retired army colonel and his wife. It was several years before their mother rejoined her children in England. The period of familial co-habitation proved short. Turing was sent to boarding school at thirteen.

At school, Turing befriended classmate Christopher Morcom. The two shared a deep interest in science and mathematics. They passed notes discussing puzzles and proofs back and forth in class. Turing came to worship the ground upon which Morcom walked. Tragically, Morcom died from tuberculosis in 1930. Turing was deeply affected by the loss.

In her memoirs, Turing's mother, Ethel Sara Turing (née Stoney), recalled her adult son with fondness:[32]

> He could be abstracted and dreamy, absorbed in his own thoughts, which on occasion made him seem unsociable [...]. There were times when his shyness led him into extreme gaucherie.

Some did not share his mother's sympathy and saw him as a loner. One of his lecturers was to speculate that Turing's isolation and his insistence on working things out from first principles bestowed a rare freshness on his work. Perhaps because of his brilliance, Turing did not tolerate fools lightly. He was also prone to eccentricity, practicing his lectures in front of his teddy bear, Porgy, and chaining his mug to a radiator to prevent theft. Turing was a rare combination of difficult to get on with but well-liked by many of his peers.

Turing won a scholarship to study at the University of Cambridge and graduated with a first-class honour's degree in Mathematics. In

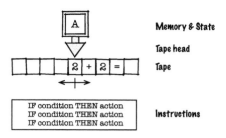

Figure 3.3 Turing machine.

the course of further studies at the university, Turing originated three important ideas in a remarkable scientific paper: he formally defined an algorithm; he defined the capabilities that a general-purpose computer must possess; and he employed these definitions to prove that some functions are not computable. Amazingly, Turing did all of this before a single digital computer was ever built.

Turing proposed an imaginary computer, now known as a Turing machine, which consists of an infinite paper tape, a tape head, a memory, and a set of instructions governing the operation of the machine. The tape is divided into cells. Each cell has enough space for just one symbol to be written. The tape head can read or write one symbol at a time to the cell directly beneath it. The machine can move the tape back and forth, cell by cell. It can also store one value in memory. The value stored is referred to as the current *state* of the machine.

Associated with the machine is a set of instructions, which control the action of the machine. The way the instructions function is quite different from how instructions work in a modern computer. In a Turing machine, every instruction is made up of a pre-condition and three actions, and the actions are performed if the pre-condition is met. The pre-condition depends on the current state of the machine (the value in memory) and the symbol currently underneath the tape head. If the state and symbol match the values specified in a pre-condition, then the actions associated with the pre-condition are performed. The permissible actions are as follows.

1. The tape head can replace, erase, or leave the symbol directly beneath it unchanged.

2. The tape can be moved left or right by a single cell, or remain still.
3. The state in memory can be updated or left unchanged

Turing envisaged that a human programmer would write programs for the machine to execute. The programmer would provide the program and input data to an associate who would manually operate the machine. With hindsight, it is easy to see that processing the instructions is so straightforward that the human operator could be replaced by a mechanical or electronic device. The operator performs the following algorithm:

Write the input values as symbols on the paper tape.
Set the initial state of the machine.
Repeat the following steps:
    Check the symbol currently under the tape head.
    Check the current state of the machine.
    Search the instructions to find a matching pre-condition.
    Perform the three actions associated with the matching
        pre-condition.
Stop repeating when the memory holds the designated halt
    state.

When the machine halts, the results are found on the paper tape.

The machine seems antiquated to modern eyes, but all of the necessary features of a computer are there—reading and writing data, executing easily modified instructions, processing symbols representing information, making decisions based on data, and repeating instructions. Turing never intended that his machine be built. Rather, it was always meant to be an abstract model of a computing machine—a conceit that would enable development of the theory of computation.

Crucially, the Turing machine manipulates symbols. It is up to humans to ascribe meaning to the symbols. The symbols can be interpreted as representing numbers, letters, logical (true/false) values, colours, or any one of a myriad of other quantities.

The Turing machine does not possess dedicated instructions for arithmetic (i.e. additions, subtractions, multiplications, and divisions). Arithmetic operations are implemented by executing programs that process the symbols on the tape so as to achieve the effect of arithmetic. For example, the calculation 2 plus 2 is performed by a program that replaces the symbols '2+2' on the tape with the symbol '4'. In a modern computer, arithmetic operations are built-in so as to increase processing speed.

Turing proposed that his machine was flexible enough to perform any algorithm. His proposal, which is now generally accepted, is a two-sided coin. It defines what an algorithm, and a general-purpose computer, are. An algorithm is a sequence of steps that a Turing machine can be programmed to perform. A general-purpose computer is any machine that can execute programs equivalent to those which can be performed by a Turing machine.

Nowadays, the mark of a general-purpose computer is that it is *Turing complete*. In other words, it can mimic the operation of a Turing machine. The paper tape symbols can, of course, be substituted by other physical quantities, e.g. the electronic voltage levels used in modern computers. All modern computers are Turing complete. If they weren't Turing complete, they wouldn't be general-purpose computers.

An essential feature of the Turing machine is its ability to inspect data and to make decisions about what action to perform next. It is this capability that raises the computer above the automatic calculator. A calculator cannot make decisions. It can process data, but not respond to it. Decision-making capability gives computers the power to perform algorithms.

Turing used his hypothetical machine to assist him in tackling a classic problem in computability: 'Can all functions be calculated using algorithms?' A function takes an input and produces a value as output. Multiplication is a computable function, meaning that there is a known algorithm for calculating the result of a multiplication for all possible input values. The question that Turing was wrestling with was: 'Are all functions computable?'.

He proved that the answer is 'no'. There are some functions that are not computable by means of an algorithm. He demonstrated that one particular function is not computable.

The halting problem queries if there is an algorithm that can always determine if another algorithm will terminate? The halting problem expresses a practical difficulty in programming. By mistake, programmers can easily write a program in which some of the steps repeat forever. This circumstance is called an *infinite loop*. Typically, it is not desirable that a program never terminates. It would be helpful for programmers to have a checker program that would analyse a newly written program and determine if it contains any infinite loops. In this way, execution of infinite loops could be avoided.

Turing proved that no general checker algorithm can exist. His proof hinges on a paradox. For which, the Liar's paradox is a good example. It is encapsulated by the declaration:

This sentence is false.

This sentence, like any logical statement, can be either true or false. If the sentence is true, then we have to conclude that: 'This sentence is false'. A statement cannot be both true and false at the same time. It is self-contradictory. Alternatively, if the sentence is false then we have to conclude that: 'This sentence is not false'—another contradiction. Since both possibilities (true-and-false as well as false-and-true) are contradictions, the statement is a paradox.

Turing used a paradox to prove that a solution to the halting problem does not exist. Paradox-based proofs work as follows:

Take a statement that you wish to prove false.
For the moment, assume that the statement is true.
Develop a logical line of reasoning based the assumption.
If the conclusion is a paradox and the logical line of reasoning is
    correct,
then the assumption must be invalid.

Turing started with the assumption that:

A checker algorithm that solves the halting problem does exist.

This hypothetical checker algorithm indicates whether a program halts or not. If the program under test halts then the checker outputs, 'Halts'. Otherwise, the Checker outputs, "Does not halt". The checker algorithm is then:

Take a program as an input.
If the program always halts,
then output "Halts",
else output "Does not halt".

Next, Turing followed a logical line of reasoning. He started by creating a program that runs the checker on itself (Figure 3.4):

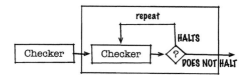

Figure 3.4 Turing's answer to the halting problem.

Repeat the following step:
    Run the checker taking the checker as input.
Stop repeating if the checker output is "Does not halt".

The program examines the output of the checker. If the checker outputs "Does not halt", the loop terminates and the program halts. If the checker outputs "Halts", the program repeats indefinitely and does not halt. However, the checker is checking itself. Thus, the checker only halts if the checker does not halt. Conversely, the checker does not halt if the checker does halt. Both outcomes are contradictions. There is a paradox.

Since the logical line of reasoning is correct, the original assumption must be invalid. This means that a checker algorithm that solves the halting problem cannot exist. The halting problem is not computable. There exist functions that, even with complete knowledge, are not computable. There is a limit to computation.

The good news is that many useful functions—mappings from input values to output values—are computable. The difficulty is coming up with efficient algorithms that will perform the desired mapping.

In 1936, Turing moved to Princeton University to undertake a PhD degree. Whereas a Bachelor of Arts, or Science, degree involves a set of taught courses and a series of written examinations, the more advanced Doctor of Philosophy degree consists of a monolithic research project. A PhD culminates in submission of a doctoral thesis and an intense oral, or *viva voce*, examination. The award of a PhD hinges on the candidate producing a novel piece of work supported by a proof, be it experimental or mathematical. Turing's PhD, supervised by the American mathematician and logician Alonzo Church, focused on the theoretical problems of computation.

Turing returned to Cambridge University in 1938. On 4 September 1939, the day after the British Declaration of War on Nazi Germany,

Turing joined the Government Codes and Ciphers School at Bletchley Park. Codenamed Station X, Bletchley Park was the UK's top-secret wartime code-breaking centre.

Based on intelligence received from a Polish group, Turing and Gordon Welchman developed a special-purpose computer to assist in decrypting the German Enigma codes. The Bombe—an electromechanical computing device—was completed in 1940. Although it could be reconfigured, the Bombe was not programmable. It, like other devices of its generation, could only perform a fixed algorithm. The Bombe allowed the team to decode intercepted radio communications from German submarines. The intelligence gleaned enabled the British Navy to determine German U-boat positions and forewarn allied shipping of imminent attack. As a direct consequence, the lives of many allied seamen were saved. Turing was awarded the Order of the British Empire in recognition of his wartime service. Details of his top-secret activities were not disclosed.

After the war, Station X was disbanded. Turing joined the National Physical Laboratory (NPL) in London, where he embarked on a project to design a general-purpose electronic computer. Due to the difficulty of the task and Turing's limitations in working with others, progress was slow. Frustrated, Turing left the NPL to return to Cambridge before taking a job at Manchester University. Manchester had pressed ahead with development of its own electronic computer.

Having spent his career contemplating the prospect of a computer, Turing finally had his hands on one. He took to programming the Manchester University Computer and wrote a manual for it. He authored a series of scientific papers pondering possible future applications of computers. Foreshadowing modern bioinformatics, he suggested that computers could predict how molecules might behave in biological systems. He suggested that computers could tackle problems that had hitherto required human intelligence. He speculated that by the year 2000, it would be impossible for a human interrogator, communicating by text message, to distinguish between a human subject and a computer. This so-called Turing test for artificial intelligence has yet to be passed.

In 1952, Turing reported a burglary at his home. He informed the police that a friend of his—Arnold Murray—knew the burglar. When pressed, Turing acknowledged a homosexual relationship with Murray. Criminal proceedings were brought against Turing for the crime of 'gross indecency'. The court sentenced Turing to hormone 'therapy'—a euphemism for chemical castration. Previously a keen amateur athlete and marathon runner, the 'therapy' seems to have had a negative effect on his health.

Two years later, at the age of just forty-one, Turing was discovered dead in his own bed. Toxicology tests indicated that the cause of death was cyanide poisoning. The coroner's inquest concluded that Turing died by suicide. Many have questioned that verdict. There was no note. A half-eaten apple was found beside Turing's bed, but it was never tested for cyanide. His friends testified that he was in good spirits in the days prior to his death. Some have suggested that Turing was assassinated by agents of the state seeking to prevent him from revealing wartime secrets. The speculation seems far-fetched. Turing habitually conducted chemical experiments at home. It may well be that his death was simply an accident.

The goings-on at Station X were hushed up until the 1970s. Even the families of the men and women who worked there knew nothing of their achievements at Bletchley Park. Belatedly, in 2013, Turing was granted a posthumous pardon for his 'crime' of homosexual activity by Queen Elizabeth.

Turing's legacy to computer science is immense. He defined computers and algorithms; he set limits on computation; his Turing machine remains the benchmark against which all computers are compared; passing the Turing test has become one of the all-time goals of artificial intelligence. More important than his inventions, though, are the small nuggets of ideas that he scattered throughout his papers. His seemingly offhand musings opened up whole avenues of future enquiry for those following in his wake. In recognition of his achievements, the greatest honour in computer science is the ACM Turing Award. The annual one million dollar prize is offered by the Association of Computing Machines (ACM) for contributions 'of lasting and major technical importance'.

While Turing was busy cracking the Enigma codes at Bletchley Park, programmable and (mostly) electronic computers were under development elsewhere. In Berlin, the German engineer Konrad Zuse built a series of relay-based electromechanical computing devices. The programmable and fully automatic Z3 was completed in 1941. Although not Turing complete, the Z3 contained many advanced features, which would later be re-invented elsewhere. Zuse's efforts were greatly hindered by the war. Lack of parts, limited funds, and aerial bombing all played their part in slowing development of the Z series. By 1945, Zuse's work was practically at a standstill. Development of the Turing complete Z4 computer was fatefully stalled. It would be 1949 before Zuse re-established a company to manufacture his devices. The Z4 was finally delivered to ETH Zurich in 1950. Zuse's company built over 200 computers before it was acquired by German electronics conglomerate Siemens.

In the US, the Second World War was a boon to the development of the first computers. Inspired by a demonstration model of Babbage's Difference Engine, Howard Aiken designed an electronic computer at Harvard University. Funded and built by IBM, the Harvard Mark I (AKA the Automatic Sequence Controlled Calculator) was delivered in 1944. Fully programmable and automatic, the machine could run for days without intervention. However, it lacked decision-making capability, and as such, the Harvard Mark I was not Turing Complete and, hence, not a general-purpose computer.

The world's first operational digital general-purpose computer was constructed in Pennsylvania. Bankrolled by the US Army, and with a clear military application, the machine was unveiled to the assembled press in 1946. It was the beginning of a revolution in algorithm development.

4

Weather Forecasts

For six days and seven nights
The wind blew, flood and tempest overwhelmed the land;
When the seventh day arrived the tempest, flood and onslaught,
Which had struggled like a woman in labour, blew themselves out.
The sea became calm, the *imhullu*-wind grew quiet, the flood held back.

> Unknown author, translated by Stephanie Dalley
> *The Epic of Gilgamesh, circa* 2,000 BCE [2]

Since time immemorial, lives have depended on the vagaries of the weather. Frequently, human catastrophe could have been averted, if only the weather were known a day in advance. In 2,000 BCE, foretelling the weather accurately was the preserve of the gods. Seafarers and farmers looked to weather lore and omens for guidance.

About 650 BCE, the Babylonians attempted to forecast the weather more precisely based on observation of cloud formations. Around 340 BCE, the Greek philosopher, Aristotle, wrote *Meteorologica*, the first significant book on the nature of weather. His immediate successor at the Lyceum compiled an accompanying text, *Theophrastus of Eresus on Winds and On Weather Signs*, which documented weather lore. These two books remained the definitive statements on the subject for almost two thousand years. Unfortunately, both were fundamentally wrong.

In the post-Enlightenment years, scientists took to painstakingly recording measurements of weather conditions. Several countries established centralized metrological services. The UK founded the Meteorological Office in 1854 and, six years later, the US Weather Bureau commenced operations. Facilitated by the recent invention of the electrical telegraph, these services collected meteorological data from outlying regions and, based on this information, issued weather forecasts. Their forecasting method was basic. Meteorologists simply searched the historical record for the closest approximation to current conditions. They then predicted that the weather would develop in the same manner once again. Sometimes, this approach worked. On occasion, it was disastrously incorrect.

The first inkling of a more accurate forecasting method came at the beginning of the twentieth century. Cleveland Abbe, Head of the US Weather Bureau, pointed out that the Earth's atmosphere is essentially a mixture of gases. He contended that gases must behave in the same way in the atmosphere, as in the laboratory. Contemporary scientists could predict what gases in the laboratory would do when subjected to heat, pressure, and movement. Surely meteorologists could use these same scientific laws to predict what gases in the atmosphere would do when subjected to the heat of the sun and the flow of the wind? The laws of hydrodynamics (forces acting on fluids) and thermodynamics (heat acting on fluids) were known. Why not apply them to the atmosphere?

Shortly thereafter, Norwegian scientist Vilhelm Bjerknes suggested a two-step method for forecasting. In the first step, the current state of the weather is measured. In the second, a set of equations are used to predict the future pressure, temperature, density, humidity, and wind velocity in the atmosphere. Bjerknes derived these equations from the known laws of physics. Due to their complexity, he was unable to wrangle a straightforward solution from the equations. Instead, he resorted to charts and graphs to convert observations to estimates of future conditions. Each chart was designed to advance the weather by a fixed time unit. By repeating these steps, Bjerknes argued, weather predictions could be projected into the future. The conditions output from the first iteration could be the input to the second and so on.

While Bjerknes' iterative method was a breakthrough, the accuracy of his approach was limited by the charts and graphs. A reliable method for calculating future conditions from current measurements was needed. The problem was fraught with difficulty. The equations were complex and seemingly impossible to solve.

A man with little previous experience of meteorology decided to take up the challenge.

Numerical Forecasts

Born in Newcastle, England in 1881, Lewis Fry Richardson (Figure 4.1) studied science at Newcastle University and King's College, Cambridge.

Figure 4.1 Numerical weather forecaster Lewis Fry Richardson, 1931. (© *National Portrait Gallery, London.*)

After graduation, he held a series of short-term research positions working on fluid flow and differential problems. In 1913, he was appointed superintendent of the remote Eskdalemuir weather observatory. Located in the uplands of southern Scotland, the Eskdalemuir landscape is beautiful, wind-swept, and stark. Richardson's duties comprised of recording the weather, monitoring seismic vibrations, and noting variations in the Earth's magnetic field. Thankfully, the job came with a house and plenty of spare time. Amidst 'the bleak and humid solitude of Eskdalemuir', Richardson embarked on a mission to develop and test a numerical algorithm for weather forecasting. Richardson's experiment was underpinned by the inviolable laws of physics.

Richardson divided the atmosphere into an imaginary three-dimensional grid of *cells*. A cell might be 100 miles wide by 100 miles long and two miles high. He assumed that within a cell, the atmosphere is relatively homogeneous. In other words, all points in the cell possess approximately the same wind speed, wind direction, humidity, pressure, temperature, and density. To capture the state of the atmosphere, he wrote down the values of these quantities for every cell. Thus, the prevailing conditions were represented as a list of numbers.

To determine how conditions evolved over time, Richardson divided every day into a series of time steps. A time step might be one

hour. Beginning with the observed conditions, he calculated the likely
state of the weather in the next time step. To make this determination, he adapted the laboratory-derived gas equations so that he could
calculate the state of a cell based on its own and its neighbours' state
in the previous time step. Richardson performed these calculations for
every cell. The completed table of values for one time step was then
used as the input for calculation of the conditions in the next time step.
Special equations were used for boundary cells at the top and bottom of
the atmosphere. The heating effects of the sun were incorporated based
on the time of day. Even the effects of the rotation of the Earth were
included. Cell by cell, time step after time step, Richardson calculated
the weather. His algorithm can be summarized as follows:

Measure the initial conditions in very cell.
Repeat the following for every time step:
    Repeat the following for every cell:
        Calculate the cell's state from its and its neighbours'
            state in the previous time step.
    Stop repeating when all cells have been processed.
Stop repeating when all time steps have been processed.
Output the completed forecast.

Today, Richardson's algorithm would be called a *simulation*. It predicts,
by means of calculation, how a real-world physical system changes
over time. The equations governing the simulation are a *model* of the
dynamics of the real-world conditions.

Richardson tested his algorithm using historical weather data measured over Germany on 20 May 1910. He sought to predict the pressure
and wind speed at two points on the map based on measurements
taken six hours earlier. Richardson rigorously applied his algorithm,
performing every calculation manually. It took months.

In the end, Richardson's predictions were horribly inaccurate. His
algorithm estimated that the surface pressure would rise to 145 hPa—
a completely unrealistic value. In fact, pressure hardly changed at all
that day. Richardson blamed the discrepancy on errors in how he
represented the initial winds.

Undaunted, Richardson published his findings in 1922 in *Weather
Prediction by Numerical Process*. In the book, he envisaged a great hall in
which 64,000 human computers aided by mechanical calculators would
compute the weather forecast in real-time.

Richardson's book was not well received. His algorithm was wildly inaccurate and outlandishly impractical. It necessitated a vast amount of computation. The only way forward was with the assistance of a high-speed calculating machine. It would be almost thirty years before numerical weather forecasting was revisited.

ENIAC

The first operational general-purpose computer was designed and built at the University of Pennsylvania during the Second World War. The ENIAC (Electronic Numerical Integrator And Computer) was designed by John Mauchly and Presper Eckert, two professors at the college. In a twist of fate, most of the credit went to world-famous mathematician, John von Neumann.

Mauchly was born in Cincinnati in 1907. Such was his ability, he was allowed to begin his PhD studies in Physics before completing the more basic BSc degree. On graduation, Mauchly was appointed as a Lecturer at Ursinus College in Pennsylvania.

In 1941, Mauchly took a course on Electronic Engineering at the Moore School in the University of Pennsylvania. Sponsored by the US Navy, the course focused on electronics for the military. Eckert, a recent graduate of the Moore School, was one of the instructors on the course. Although Eckert hadn't been an ace student, he was a superb practical engineer. Despite Mauchly being Eckert's senior by twelve years, the two hit it off, bonding over a shared fascination with gadgets. After the course, Mauchly was hired by the Moore School.

In the shadow of the Second World War, the Moore School was host to human computers working for the US Army. These human computers were employed by the Ballistic Research Laboratory (BRL) situated at the nearby Aberdeen Proving Ground in Maryland. The team, assisted by the Moore School's mechanical calculators, produced ballistics tables for the artillery. The tables were used by gunnery officers on the battlefield to determine the correct firing angles for their artillery pieces. The tables allowed an officer to take into account the equipment type, air pressure, wind velocity, wind direction, target range, and target altitude. The BRL employed one hundred female graduate mathematicians to perform the exacting and time-consuming calculations. Even with that workforce, the Laboratory was unable to keep up with demand.

In April 1942, Mauchly wrote a proposal outlining the design of an electronic computer. His report was circulated within the Moore School. Lieutenant Herman Goldstine, the leader of the trajectory tabulation effort, heard about the document. A Professor of Mathematics at the University of Michigan, Goldstine immediately saw the potential for Mauchly's computer to alleviate the bottleneck in the ballistics calculations. He contacted Mauchly. Pleased with what he heard, Goldstine applied to his Army superiors for funding to enable Mauchly and Eckert to build the machine.

Work started on the ENIAC in 1943. The youthful and energetic Eckert was appointed Chief Engineer. The mature Mauchly acted as Consultant to Eckert. Goldstine was Project Manager *cum* Mathematician.

Although intended for computation of artillery tables, ENIAC was a fully-fledged general-purpose computer. It was programmable, albeit with wires and plugs. It could perform calculations, store values, and make decisions. Like Babbage's Analytic Engine, ENIAC processed decimal numbers. Mostly electronic, with some electromechanical components, the machine was immense, weighing roughly twenty-seven tonnes and covering over 1,500 square feet of floor space. Racks of electronic circuits were set into floor-to-ceiling cabinets lining the walls of the Moore School basement. Row upon row of light bulbs and plug sockets adorned the front of the cabinets. More cabinets, this time on wheels, shuttled between the banks of equipment. Multitudes of cables snaked between sockets in seemingly incomprehensible patterns.

Amidst the electronic equipment, a small cadre of programmers toiled to cajole the temperamental machine into action. The ENIAC programmers (Figure 4.2) where taken from the ranks of the mathematicians working for the BRL. Kathleen McNulty (later Mauchly and Antonelli), (Betty) Jean Jennings (later Bartik), (Frances) Betty Holberton (later Snyder), Marlyn Wescoff (later Meltzer), Frances Bilas (later Spence), and Ruth Lichterman (later Teitelbaum) figured out how to program the bewildering machine. It was a nightmare. Multiple units operated simultaneously. The outputs of each had to be re-timed so as to synchronize the data transfer. Components kept breaking down. To make the task somewhat easier, Adele Goldstine (née Katz), Goldstine's wife and an instructor at the School, compiled the first operating manual for the machine.

As ENIAC neared completion, Mauchly and Eckert began to contemplate its successor. In August 1944, they proposed an improved

Figure 4.2 The ENIAC team, 1946. Left to right: Homer Spence; J. Presper Eckert, chief engineer; Dr John W. Mauchly, consulting engineer; Elizabeth Jennings (aka Betty Jean Jennings Bartik); Capt. Herman H. Goldstine, liaison officer; Ruth Lichterman.

device—the EDVAC (Electronic Discrete Variable Automatic Computer). Again, BRL approved funding. Work on EDVAC commenced shortly thereafter. At roughly the same time, Goldstine introduced a new collaborator to the project.

John von Neumann was born in Budapest, Hungary, in 1903. From a well-to-do family, he was privately educated until the age of ten. At secondary school, he showed a special aptitude for mathematics, to the extent that he wrote his first research paper before he was eighteen. Von Neumann went on to study mathematics at the University of Budapest. At the same time, he completed a Chemistry degree in Zurich, Switzerland. He didn't bother attending the course in Zurich. He just turned up for the exams. He would come to be fluent in English, German, and French, with passable Latin and Greek.

Von Neumann was appointed Professor at Princeton University in 1931. There, he joined Einstein as a member of the Institute for Advanced Study (IAS). Leon Harmon, who worked at the IAS, described von Neumann as:[45]

> a true genius, the only one I've ever known. I've met Einstein and Oppenheimer and Teller and a whole bunch of those other guys. Von Neumann

was the only genius I ever met. The others were super-smart and great prima donnas. But von Neumann's mind was all-encompassing. He could solve problems in any domain and his mind was always working, always restless.

With a warm and friendly personality, von Neumann was well liked. Everyone knew him as 'Johnny'. He had the humility to listen and the ability to absorb ideas. Given to sharp suits and fast cars, Johnny was possessed of an earthy sense of humour. The great intellectual delighted in people and gossip.

During the Second World War, von Neumann was granted a leave of absence from Princeton to contribute to military projects. He was heavily involved in the Manhattan Project, assisting in the design of the first atomic bomb. The Project demanded a great many calculations. Von Neumann saw the need for a machine that could calculate faster than any human. In 1944, von Neumann met Goldstine, by chance it seems, on a train station platform in Aberdeen, Maryland. Goldstine introduced himself. The two got to talking and, perhaps to impress von Neumann, Goldstine mentioned his work on ENIAC. Von Neumann's interest was piqued. Goldstine extended an invitation and, subsequently, von Neumann joined the ENIAC project as a consultant. Eckert later said: [47]

Von Neumann grasped what we were doing quite quickly.

In June 1945, von Neumann wrote a 101-page report entitled *First Draft of a Report on the EDVAC*. The report described the new EDVAC design in detail but neglected to mention Mauchly and Eckert, the machine's inventors. With Goldstine's approval, the report was distributed to people associated with the project. Von Neumann, sole author of the first report on EDVAC, was widely seen as the originator of the design. Eckert complained: [47]

I didn't know he was going to go out and more or less claim it as his own. He not only did that, but he did it at the time when the material was classified, and I was not allowed to go out and make speeches about it.

Completed in 1945, ENIAC arrived too late to assist in the war effort. The giant machine was unveiled to the public on St Valentine's Day 1946 at a press conference in the Moore School. One of the team, Arthur Burks, gave a demonstration of ENIAC's capabilities. He started the show by adding 5,000 numbers together in one second. Next, Burks explained that an artillery shell takes thirty seconds to travel from gun to target. In contrast, manual calculation of its trajectory takes three days. He informed the audience that the ENIAC would now perform the same computation. With a touch of theatre, the main lights were switched off so that the assembled reporters could more clearly observe the machine's flickering lights. Twenty seconds later—faster than the shell could fly—the calculation was complete.

That evening, the project luminaries gathered for a celebratory dinner. The dinner was for the top brass and electronic engineers only. ENIAC's female programmers were not invited.[50] It would be fifty years before the ENIAC programmers received a modicum of the recognition that they so richly deserved.

The morning headlines were euphoric:[51]

ARMY'S NEW WONDER BRAIN AND ITS INVENTORS
ELECTRONIC 'BRAIN' COMPUTES 100-YEAR PROBLEM
IN 2 HOURS

ENIAC was moved to its owner's premises, the BRL facility in Aberdeen, in 1947. The machine remained in use until 1955. Disgruntled, Mauchly and Eckert, resigned from the Moore School in 1947 to found their own computer company. The nascent corporation soon ran into financial distress and was acquired by Remington Rand in 1950.

Controversially, Eckert and Mauchly's patent application for the ENIAC was disallowed. The judge's decision rested, in part, on the prior existence, and Mauchly's knowledge of, the Atanasoff-Berry Computer (ABC). Developed by John Atanasoff, a professor at Iowa State University, and his student, Clifford Berry, the ABC was electronic. However, the ABC was not programmable and lacked decision-making capability. By modern standards, the ABC was certainly not a general-purpose computer. It was a special-purpose electronic calculator. With hindsight, Judge Larson's ruling is perplexing. ENIAC was far more advanced than the ABC and contained many innovative features.

Von Neumann's paper stripped Mauchly and Eckert of fame. The failure of their start-up and patent application deprived them of fortune. After a career in computing, Mauchly passed away in 1980. Eckert, the younger of the pair, worked at Remington Rand and its successors, until he died in 1995.

Although ENIAC was designed for ballistics calculations, it seems that the first operational runs were for a higher purpose. The first runs were secret calculations for the Manhattan Project. At von Neumann's suggestion, a group from Los Alamos visited ENIAC in 1945. Impressed, they lobbied for ENIAC to be made available to aid in the computations needed for design of a hydrogen bomb. Their requests were granted, and an ongoing relationship was established between the Manhattan Project and the ENIAC team. This collaboration allowed testing of one of the most powerful algorithms in history.

Monte Carlo

Stanislaw Ulam (Figure 4.3) was born to a well-off Polish-Jewish family in 1909. He studied mathematics, graduating with a PhD from the Lviv Polytechnic Institute in the Ukraine. In 1935, he met John von Neumann in Warsaw. Von Neumann invited Ulam to work with him for a few months at the IAS in Princeton. Soon after joining von Neumann, Ulam procured a lecturing job at Harvard University in Boston. He moved permanently to the US in 1939, narrowly avoiding the outbreak of the Second World War in Europe. Two years later, he became a citizen of the United States. Ulam forged a reputation as a talented mathematician and, in 1943, was invited to join the Manhattan Project in Los Alamos, New Mexico. The high-powered, collaborative environment at Los Alamos suited Ulam. Nicholas Metropolis, a Los Alamos colleague, later wrote of Ulam:[52]

> His was an informal nature; he would drop in casually, without the usual amenities. He preferred to chat, more or less at leisure, rather than to dissertate. Topics would range over mathematics, physics, world events, local news, games of chance, quotes from the classics—all treated somewhat episodically but always with a meaningful point. His was a mind ready to provide a critical link.

In the push to build the bomb, Ulam was assigned the problem of calculating the distance that neutrons (charge-free particles at the centre of an atom) travel through a shielding material. The problem seemed intractable. Neutron penetration depends on the particle's trajectory and the arrangement of atoms in the shielding material. Imagine a table tennis ball carelessly launched at a million skittles placed at random. How far does the ball travel, on average? There are so many possible paths, how could anyone answer the question?

While in hospital convalescing from an illness, Ulam took to playing Canfield Solitaire, a single-player card game. Canfield Solitaire uses the normal deck of fifty-two playing cards. Cards are dealt one by one and moved between piles according to the rules of the game and the player's

Figure 4.3 Stanislaw Ulam, inventor of the Monte Carlo method, *c*. 1945. (*By Los Alamos National laboratory. See Permissions.*)

decisions. The goal is to end up with just four piles of cards. Each pile should contain all of the cards from one suit.

The rules are quite simple. When a card is drawn, there is a small number of legal moves to choose from. In most cases, selecting the best move is straight-forward.

Ulam wondered, what were his chances of winning a game? Whether he won or lost depended on the order in which the cards were dealt. Some sequences of cards lead to a win, others a loss. One way to calculate the odds was to list all possible card sequences and count the percentage that lead to wins.

Since a deck contains fifty-two cards, there are fifty-two possible first cards (Figure 4.4). After that, there are fifty-one cards in the pack, so there are fifty-one possible second cards. Thus, the number of possible first and second card sequences is $52 \times 51 = 2,652$. Extending this calculation to the whole pack, gives $52 \times 51 \times 50 \times 49 \times \cdots \times 1$. This is equal to an 8 followed by sixty-seven digits. No one could possibly play that many games.

Ulam wondered if the problem could be simplified. What if he played just ten games? He could count the percentage of wins. That would give him an indication of the true probability of winning. Of course, with just ten games there is the possibility of a lucky streak. This would distort the odds. What about one hundred games? A lucky streak that long

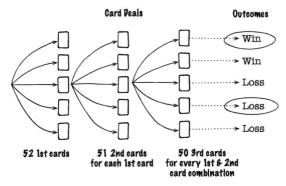

Figure 4.4 Possible card deals in Canfield Solitaire. The circled outcomes are sampled using the Monte Carlo method.

is less likely. Ulam concluded that he could get a reasonable estimate of the true probability of winning by averaging the outcomes over a sufficiently large number of games. The key point was that he did not have to play all possible games. He just had to play enough to get a reasonable estimate of the true odds.

Still, even one thousand games would take a long time to play. Ulam realized that a computer could be programmed to play that many games. The computer could be programmed to 'deal' the cards randomly and play in the same way as Ulam. With enough games, the percentage of wins would be a reliable estimate for the true winning probability.

In summary, Ulam's algorithm operates as follows:

Set the win count at zero.
Repeat the following:
    Take a fresh pack of cards.
    Repeat the following:
        Draw a card at random.
        Play the card in the best way.
    Stop repeating when the pack is empty.
    If the game was won, then add one to the win count.
Stop repeating after a large number of games.
Output the percentage of wins.

Ulam figured that his algorithm would work for more than just card games. It would also work for the neutron diffusion problem.

The neutron trajectories and the shielding atom positions could be represented by random numbers. The penetration distance could be calculated for each trajectory and shielding configuration. Averaging over a large number of random trials would give an estimate of the typical real-world neutron penetration distance.

Ulam proposed the idea to von Neumann and suggested that the neutron penetration calculations be run on ENIAC. According to colleague Nicholas Metropolis:[55]

> Von Neumann's interest in the method was contagious and inspiring. His seemingly relaxed attitude belied an intense interest and a well-disguised impatient drive.

ENIAC was quickly tasked with testing Ulam's new method. The results were considered 'quite favourable'—a polite understatement. The card players back at Los Alamos named the new algorithm the Monte Carlo method after the famous casino in Monaco.

Metropolis and Ulam published the first paper on the Monte Carlo method in 1949. The method has gone on to become a staple of computer simulation. It allows scientists to estimate the probable outcomes of complex physical events by randomly sampling a large number of cases. Today, the Monte Carlo method is intrinsic to speculative studies in fields as diverse as physics, biology, chemistry, engineering, economics, business, and law.

Computer Forecasts

After the War, von Neumann returned to academic life at the Institute for Advanced Study in Princeton (Figure 4.5). There, he initiated a project to build a new electronic computer along the lines of the EDVAC. The IAS computer, as it became known, was to be von Neumann's gift to computing. The machine was operational from 1952 until 1958. More importantly, von Neumann distributed plans for the IAS machine to an array of research groups and corporations. The IAS machine became the blueprint for computers all over the world.

In addition, Von Neumann pondered the tasks that computers might be put to. Perhaps as a result of his work on fluid flow in the 1930s, von Neumann seems to have been aware of Richardson's work on numerical weather forecasts. Inspired, von Neumann secured a grant from the US Navy for establishment of the first computer meteorology research group. To kick start the initiative, von Neumann organized a conference

Figure 4.5 Mathematician John von Neumann with the IAS computer, 1952. (*Photograph by Alan Richards. Courtesy of the Shelby White and Leon Levy Archives Center at the Institute for Advanced Study, Princeton, NJ, USA.*)

to bring together the leading meteorological researchers. The group took on the challenge of conducting the first weather forecast performed by a computer.

One of the group, Jule Charney, joined von Neumann at the IAS. Charney tamed the complexity of the gas equations, rendering them amenable to computer execution. The IAS computer wasn't quite ready, so von Neumann again requested time on ENIAC.

On the first Sunday of March 1950, five meteorologists arrived at the BRL in Aberdeen to perform the forecast. The meteorology and programming teams worked around the clock, in eight-hour shifts, for almost the entirety of the next thirty-three days. Forecasts were computed for North America and Europe on four days in January and February 1949. The dates were specially selected due to the presence of significant weather systems. The forecasts predicted barometric pressure over twenty-four hours. Data from the US Weather Bureau provided the initial conditions and allowed evaluation of the resulting forecasts. The model was based on a rectangular grid of fifteen by eighteen cells, each 736 km across, with one-hour time steps. ENIAC performed the one million calculations needed for each forecast in about twenty-four hours—just keeping pace with the unfolding weather.

The results were mixed. Some features were predicted accurately. Others, such as the position and shape of a cyclone, were incorrect. The researchers put the errors down to the large cell sizes and the limitations of the equations. Nonetheless, the concept of numerical weather prediction by computer was proven. The meteorologists and programmers just had to work out the details.

Richardson was vindicated at last. Charney mailed a copy of the final paper describing the 1950 ENIAC forecasting experiments to Richardson. The pioneer of numerical forecasting replied, congratulating the team. In a self-deprecating aside, Richardson remarked that the outcome of the ENIAC experiments was an:[40]

> ...enormous scientific advance on [his own] single, and quite wrong result.

Richardson passed away in 1953, just two years later.

John von Neumann died in 1957 at the age of fifty-three after a long battle with cancer. Hans Bethe, a Nobel Laureate, remarked:[60]

> I have sometimes wondered whether a brain like von Neumann's does not indicate a species superior to that of man.

Von Neumann's obituary was written by Stanislaw Ulam.

Ulam went on to make significant contributions to nuclear physics, bioinformatics, and mathematics. He held professorial appointments at a series of prestigious American universities, while working summers at Los Alamos. Ulam passed away in Santa Fe, New Mexico in 1984.

As a result of refinements in algorithms, advances in computer performance, and increases in the number of weather monitoring stations, forecasting accuracy steadily improved during the 1950s and 1960s. All seemed to be progressing well until Edward Lorenz stumbled upon a fundamental limitation of Richardson's approach.

Chaos

Edward Lorenz was born in Connecticut in 1917. He studied mathematics prior to serving as a meteorologist in the US Army Air Corps. He went on to study the subject at Massachusetts Institute of Technology (MIT) before becoming a professor there. His discovery of a problem with Richardson's method happened by chance in 1961.

As part of a research project, Lorenz ran a series of weather simulations on a small computer. The exercise should have been routine. What transpired was distinctly odd.

Lorenz elected to repeat one of the simulations to examine the results in greater detail. On returning to the computer an hour later, he compared the new results with the previous outputs. He was startled to find that the new forecasts were nothing like the old ones. Naturally, he suspected that the cause was a fault in the computer—a common occurrence at the time. Before calling computer maintenance, he decided to check the simulation print-outs time step by time step. At the beginning, they matched. After a while, the values began to diverge. The discrepancy grew rapidly as the simulation evolved. The difference between old and new more or less doubled every four simulated days. By the end of the second simulated month, the old and new outputs bore no resemblance to one another whatsoever.

Lorenz concluded that both the computer and the program were working just fine. The discrepancy arose from a very small difference in the simulation inputs. On the first run, Lorenz had entered the state of the atmosphere as six digits. On the second run, he had only typed in three digits. The difference between a six-digit number and the nearest three-digit number is small. One would except a small difference at the input to lead to a small difference at the output. On the contrary, the calculations had caused the difference to grow, ultimately leading to a large discrepancy between the outputs. Crucially, Lorenz realized that he wasn't seeing an artefact of simulation. The simulation was accurately modelling a real-world phenomenon.

A new science sprang from Lorenz's accidental discovery. *Chaos theory* has since determined that many real-world physical systems are hypersensitive to their initial conditions. Small changes in the initial state can lead to major differences in conditions later on. The idea was encapsulated in the popular epithet, 'the Butterfly Effect'. If conditions are right, the flap of a butterfly's wings in Brazil can be the sole cause of a tornado in Texas a few days later. The example is extreme. Nevertheless, the idea has proven its worth. Chaos has since been identified in many real-world systems, including the orbits of asteroids moving within the rings of Saturn.

Chaos theory places a bound on the time horizon of numerical weather forecasts. Small errors in modelling current conditions can lead to large forecasting errors at a later stage. This time horizon for accurate forecasting seemed inviolable until Edward Epstein stepped in. Epstein was born in the Bronx in 1931. Like Lorenz, Epstein was introduced to meteorology during military service. After leaving the US Air Force, Epstein worked as a researcher and lecturer at a series of American colleges. While a visiting scientist at the University of Stockholm, he published a paper outlining an algorithm that could mitigate the Butterfly Effect. Epstein's idea offered a way to extend the weather forecasting time horizon.

Numerical weather forecasting, as proposed by Richardson, relies on a single simulation to predict the weather. The simulation starts with measurements of current conditions and, cell by cell, time step after time step, calculates how the weather evolves. Epstein's insight was to apply Ulam's Monte Carlo method to Richardson's numerical simulation.

Instead of one simulation, Epstein proposed that many simulations be run. Each simulation begins with randomly *perturbed* initial conditions. These initial conditions are created by applying small random changes, or perturbations, to the observed atmospheric conditions. Given the limitations of measurement, forecasters don't know exactly what the current weather conditions are in any cell. The perturbations allow forecasters to try out a number of possible scenarios. At the end of the simulations, the outputs are averaged to obtain a unified final prediction. This average takes an *ensemble* of possibilities into account. Their average is the middle-of-the-road, all-things-considered, most likely scenario. Epstein's proposal can be summarized thus:

Measure the current atmosphere conditions.
Repeat the following steps:
    Add small random perturbations to the current conditions.
    Perform a numeric forecast starting from these initial
       conditions.
    Store the results.
Stop repeating when sufficient simulations have been
   performed.
Output the average forecast.

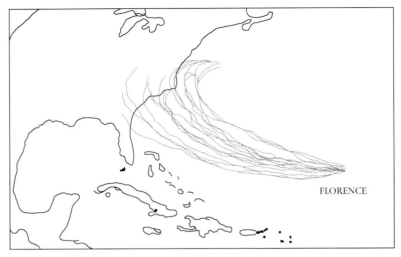

Figure 4.6 Hurricane Florence (2018) path predictions obtained using an ensemble method.

The downside of Epstein's algorithm is that it requires a lot of calculation. Running eight Monte Carlo simulations requires eight times the computer power of a single forecast. For this reason, Epstein's ensemble approach wasn't put into operational use until the early 1990s. Today, ensemble forecasts are state-of-the-art (see Figure 4.6). For example, the European Centre for Medium-Range Weather Forecasts now bases its predictions on fifty-one unique simulations.

Edward Lorenz was to accumulate a long list of scientific prizes for his work on chaos theory. Edward Epstein devoted his entire career to meteorology and climate modelling. He was an early advocate for the concept of man-made climate change. Both passed away in 2008.

Long-Range Forecasts

Like ENIAC, the computers of the 1950s were expensive, power hungry, and unreliable behemoths. The invention of the *transistor* and the *integrated circuit*, in 1947 and 1958 respectively, allowed the miniaturization of the computer.

Transistors are electronic switches. Containing no moving parts, other than electrons, transistors are small, low power, reliable, and incredibly fast. Groups of transistors can be wired together to create logic circuits. These logic circuits can, in turn, be interconnected so as to create data processing units.

Integrated circuits allow the manufacture of vast numbers of tiny transistors, together with their metallic inter-connections, at incredibly low cost. An integrated circuit is inside every *computer chip*. These chips are the physical building blocks of modern computers.

Over the years, electronic engineers have refined transistor designs and integrated circuit technology. In 1965, Gordon Moore, the co-founder of Intel, noted that his engineering team had managed to double the number of transistors fabricated on a single integrated circuit every eighteen months. Moore saw no reason why this trend could not continue into the future. His prediction, enshrined as 'Moore's Law', became a roadmap for the industry. Moore's Law has proven to be one of the greatest predictions of the modern age. The Law has held true for more than half a century.

In line with Moore's prediction, the performance of computers has risen exponentially. Even as performance has risen, computer size, cost, and power consumption have plummeted. Today, a top-end computer chip contains tens of billions of transistors. Moore's Law is the driving force behind the vertiginous rise of the computer.

In 2008, Peter and Owen Lynch (University College Dublin and IBM Ireland) repeated the original ENIAC forecasts on a mobile phone. Their program—PHONIAC—ran on an off-the-shelf Nokia 6300 mobile phone. ENIAC took twenty-four hours to perform a single forecast. In contrast, the Nokia 6300 took less than one second. Whereas ENIAC weighted twenty-seven tonnes, a Nokia 6300 weighs just ninety-one grams. This is Moore's Law in action.

Advances in computer technology have enabled new algorithms to be developed. Algorithms that were once impossibly time consuming to execute are now routine. Avenues for algorithmic research that once seemed purely theoretical are now entirely practical. The emergence of novel computing devices has enabled fresh applications and created the need for brand new algorithms. Last, but not least, the success of the computer industry has led to huge increases in the number of people engaged in software and algorithm development.

Thus, Moore's Law has also led to exponential growth in the number of algorithms.

5

Artificial Intelligence Emerges

The machine is not a thinking being, but simply an automaton
which acts according to the laws imposed upon it.

LUIGI FEDERICO MENABREA and ADA LOVELACE
Sketch of the Analytical Engine, 1843[28]

In the 1940s and 1950s, a computer was considered to be, in essence, a fast
calculator. Due its high cost and large size, a computer was a centralized,
shared resource. *Mainframe* computers churned through huge volumes
of repetitive arithmetic calculations. Human operators were employed
as gatekeepers to these new-fangled contraptions, apportioning valuable compute time to competing clients. Mainframes ran substantial
data processing *jobs* one after another with no user interaction. The
final voluminous printouts were presented in *batch* by the operators to
their grateful clients.

Amidst the expansion of industrial-scale arithmetic, a handful of
visionaries wondered if computers could do more. These few understood that computers were fundamentally symbol manipulators. The
symbols could represent any sort of information. Furthermore, they
opined that, if the symbols were manipulated correctly, a computer
might even perform tasks which had, until that point, required human
intelligence.

More Than Math

Alan Turing left the NPL in 1947 to return to Cambridge University for a
one-year sabbatical. In departing the NPL, Turing abandoned his brain-
child, the Automatic Computing Engine (ACE). The ACE was meant to
be the UK's first general-purpose computer. However, the project did
not go well. Construction of the machine was exceedingly challenging.
In addition, Turing was difficult to work with.[67] In the wake of his

Figure 5.1 AI pioneer Christopher Strachey. (© *Bodleian Library & Camphill Village Trust. Courtesy of The National Museum of Computing.*)

departure, the team pressed on. A much simplified design, the Pilot ACE, finally became operational in 1950.

That autumn, the group received an unusual request. Christopher Strachey, a teacher at Harrow School, enquired if he might have a go at programming the Pilot ACE. Strachey was undoubtedly a novice to programming, although in 1950, everyone was a novice to programming.

Born in 1916, Strachey was scion of a well-to-do, intellectual, English family. He graduated King's College, Cambridge with a degree in Physics. In his third year, he suffered a mental breakdown. Later, his sister attributed the collapse to Strachey coming to terms with his homosexuality.[68] During the Second World War, Strachey worked on radar development. Thereafter, he took up employment as a teacher at Harrow, one of the most exclusive public schools in England.

Strachey's request was approved and he spent a day of his Christmas vacation at the NPL, absorbing all the information that he possibly could about the new machine. Back at Harrow, Strachey took to writing a program for the Pilot ACE. With no machine at his disposal, Strachey wrote the program with pen and paper and tested it by imaging the computer's actions. Most beginners start with a simple programming task. Due to either ambition or naivety, Strachey embarked on writing a program to play Checkers (Draughts in the UK). This was certainly not an arithmetic exercise. Playing Checkers mandates logical reasoning and foresight. In other words, playing Checkers requires intelligence.

That spring, Strachey got wind of a new computer at Manchester University. The project was initiated by Bletchley Park alumnus Max

Newman just after the war. More powerful than the ACE Pilot, the Manchester Baby seemed better suited to Strachey's work. Strachey got in touch with Turing, who by then was Deputy Director of the Manchester Computing Machine Laboratory. Acquaintances since King's College days, Strachey managed to wheedle a copy of the programming manual from Turing. Later that summer, Strachey visited Turing to find out more.

A few months later, Strachey returned to test a program he had written at Turing's behest. Overnight, Strachey went from handwritten notes to a working thousand-line program. The program solved the problem that Turing had set and, on completion, played *The National Anthem* on the computer's sounder. This was the first music ever played by a computer. Even Turing was impressed. It was clear that Strachey was a born programmer.

Strachey was recruited by the National Research and Development Corporation (NRDC). The NRDC's remit was to transfer new technologies from government agencies to the private sector. The NDRC didn't have much for Strachey to do at the time, so he continued programming, and among other things, he invented a program to compose love letters.

Strachey's program took a template love letter as input and selected the adjectives, verbs, adverbs, and nouns at random from pre-stored lists. From whence came the ardent epistle:[72]

Honey Dear
My sympathetic affection beautifully attracts your affectionate enthusiasm. You are my loving adoration: my breathless adoration. My fellow feeling breathlessly hopes for your dear eagerness. My lovesick adoration cherishes your avid ardour.
Yours wistfully
M. U. C. [Manchester University Computer]

To the bemusement of his colleagues, Strachey pinned the love letters on the Laboratory noticeboard. While whimsical in nature, Strachey's program was the first glimmer of computer creativity.

Strachey finally completed his Checkers program in 1952, describing it in a paper entitled *Logical or Non-Mathematical Programmes*.

Checkers is a two-player board game played on the same eight-by-eight grid as Chess. Players take opposing sides of the board and are given twelve checkers (disks) each. One player plays white, the other black. To

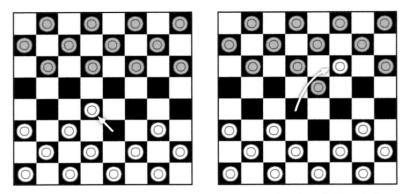

Figure 5.2 Checkers boards illustrating a simple play by white (left) and a later jump which removes a black checker (right).

begin, the checkers are placed on black squares in the three rows nearest the player (Figure 5.2). Players take turns to move a single checker. Checkers normally move one square diagonally in a forwards direction. Checkers can jump over an opponent's neighbouring checker if the square beyond is unoccupied. A sequence of such jumps can be performed in a single play. All of the opponent's 'jumped' checkers are removed from the board. The aim of the game is to eliminate all of an opponent's checkers. At the start, checkers can only move forwards. When a piece reaches the far side of the board, it is 'crowned' by placing a checker on top of it. Crowned pieces can be moved diagonally forwards or backwards.

Checkers is complex. There is no simple strategy that inevitably leads to a win. Potential plays must be evaluated by imagining how the game will evolve. A seemingly innocuous play can have unforeseen repercussions.

Strachey's algorithm uses numbers to record the position of the checkers on the board. On its own turn, the algorithm examines all possible next plays (ten on average). In board game parlance, a *move* consists of two plays, one by each player. A single play (or half-move) is called a *ply*. For every possible next play, the algorithm assesses its opponent's potential responses. This lookahead procedure is applied up to three moves deep. The results of the lookahead can be visualized as a *tree* (Figure 5.3). Every board position is a *node*, or branching point, in the tree. Every possible play from that position gives rise to a branch leading

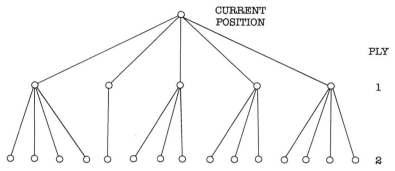

Figure 5.3 Visualization of the Checkers lookahead tree. Every node is a board position. Every branch is a play.

to the next board position. The greater the lookahead, the more layers in the tree. For the nodes at the end of the lookahead, the algorithm counts the number of checkers that each player retains on the board. It selects the play at the root of the tree that leads to the greatest numerical advantage for the computer at the end of the lookahead.

On the Ferranti Mark I, a commercial version of the Manchester Mark I, every play took one to two minutes of computer time. Even at that, Strachey's program wasn't a particularly good player. With hindsight, the program's lookahead depth was insufficient, the decision-making logic lacked sophistication, and position evaluation was inaccurate. Nevertheless, the hegemony of computer arithmetic had been broken. Here was the first working example of *artificial intelligence*.

Strachey went on to become the University of Oxford's first Professor of Computer Science. Unfortunately, while a well-regarded academic, much of Strachey's later work remains unrecognized due to his hesitancy to publish academic papers. After a short illness, he passed away in 1975, aged 58.

Board games have become the barometers of artificial intelligence (AI). The reasons why are both technical and very human. Board games have clearly defined goals and rules that are amenable to computer programming. At any point in time, there is a limited number of options available to the computer, which makes the problem tractable. Playing humans is an easily understood indictor of progress. Additionally people love contests. A machine that might beat the world champion will always provoke public interest. Even AI researchers crave an audience.

The Trouble with AI

The term 'artificial intelligence' was coined in 1955 by John McCarthy in a proposal submitted to the Rockefeller Foundation. The document requested funding for a two-month, ten-person summer research project involving luminaries such as Claude Shannon and Marvin Minsky. McCarthy, an Assistant Professor of Mathematics at Dartmouth College in New Hampshire, USA, wrote:[74]

> For the present purpose the artificial intelligence problem is taken to be that of making a machine behave in ways that would be called intelligent if a human were so behaving.

The term caught on, but McCarthy's definition has proven problematic.

There are many aspects of intelligence, but for a moment, let us consider one specific example. Before 1940, most people would have said that playing Checkers requires intelligence. The player must understand the board position and come up with a series of plays that will lead to a win. In contrast, most people would concur that performing an algorithm does not require intelligence. Slavishly carrying out one well-defined step after another is trivial. Even a machine can do that.

Therein lies the rub. Checkers requires intelligence when the algorithm for playing is unknown. As soon as the algorithm is known, playing Checkers no longer requires intelligence.

AI—as defined by McCarthy—is always the unsolved problem. Once the algorithm is known, the problem no longer requires intelligence. Intelligence, we feel, is reserved for intellectual tasks that computers cannot perform. In ways, AI is akin to stage magic. Once we know how the trick is done, it is no longer magic.

As algorithms and computers have improved, the boundary of human intelligence has repeatedly been redrawn. Before calculators, one might have said that arithmetic requires intelligence. Before Strachey, one might have claimed that playing board games needs intelligence. Looking to the future, one wonders where the ultimate boundary between algorithms and intelligence lies. Perhaps human intelligence, in its entirety, will prove to be an algorithm.

McCarthy's definition of AI has caused much confusion. The layperson imagines that AI is fully formed, equivalent to human intelligence. In truth, human intelligence is multifaceted and general—multi-faceted in that there are many aspects to our intelligence. We can learn, recall, multitask, invent, apply expertise, imagine, perceive, abstract, and so on. Human intelligence is general in that we can perform a wide variety of tasks. We can make lunch, we can debate, we can navigate, we can play sports, we can mend broken machines, and so forth. To be called 'artificially intelligent', a computer need only perform a single task that was previously thought to require human intelligence. A computer need only play Checkers to be considered in possession of AI. The proper term for intelligence akin to our own is *human-level artificial general intelligence* (HLAGI). HLAGI is what we see in science fiction movies. Science fact—AI—is very far from HLAGI.

An important aspect of McCarthy's definition was that he described AI in terms of outcome. He didn't require that machines solve problems in the same way as humans. So long as the human and machine produce comparable outcomes then he considers AI to be on a par with human intelligence. For McCarthy, the mechanism does not matter.

Over the years, the dichotomy between how computers and how humans perform tasks that require intelligence has engendered much philosophical debate. The crux of the matter can be best expressed in the simple query: 'Can machines think?' The answer, of course, depends on what is meant by the word 'think'. If thought is a biological process of the brain, then clearly computers cannot think. To most people, such a requirement seems overly restrictive. Why should the substrate delineate thinking from non-thinking? If aliens arrived from another planet, would we deny that they can think simply because they are silicon-based lifeforms, rather than carbon-based? I think not.

To most people, the prerequisites for thought are intelligence plus consciousness. Consciousness, the state of being self-aware, allows sentient beings to 'hear' themselves 'think'. For most people, consciousness is central to thinking. So far, computers are certainly not conscious. Furthermore, we have no idea how to make them conscious. Thinking machines are a long way off. Perhaps they are impossible.

McCarthy intimated that the question 'Can machines think?' is irrelevant. Alan Turing concurred:[36]

> The original question, 'Can machines think?' I believe to be too meaningless to deserve discussion. Nevertheless, I believe that at the end of the century the use of words and general educated opinion will have altered so much that one will be able to speak of machines thinking without expecting to be contradicted.

For Turing, if a machine's behaviour is indistinguishable from human intelligence then we will conclude that the machine 'thinks'. To him, whether it really 'thinks', or not, only matters to the philosopher.

In contrast, whether a machine 'feels' or not has major ramifications. If the machine has consciousness and emotion, then surely, we have ethical responsibilities towards it. Such questions will gain in import as technology advances.

McCarthy's initiative was approved for funding by the Rockefeller Foundation. The Dartmouth Conference went ahead in the summer of 1955. The gathering heralded the beginning of AI as a field of research.

Disappointingly, the Conference proved to be a meandering series of getting-to-know-you events. Participants drifted in and out of the Conference, touting their own agendas. Few gleaned new insights. Most opined that the Conference did not achieve much. [45] In retrospect, one presentation did offer a signpost to the future. Two American scientists, Allen Newell and Herbert Simon, unveiled a computer program that could perform algebra.

Machine Reasoning

Algebra is the branch of mathematics concerned with equations that include unknown values. The unknowns are signified by letters. Using the rules of algebra, mathematicians seek to re-arrange and combine equations so that the values of the unknowns can be determined. For thousands of years, manipulating equations was the domain of mathematicians. This was something that abaci, calculators, and early computer programs could not do. In 1946, John Mauchly, co-inventor of ENIAC wrote:[51]

> I might point out that a calculating machine doesn't know how to do algebra, but only arithmetic.

By the time of the Dartmouth Conference, Newell and Simon were working at RAND Corporation. Based in Santa Monica, California, RAND was, and still is, a not-for-profit research institute. Established after the Second World War, RAND specializes in planning, policy, and decision-making research for governmental bodies and corporations. In the 1950s, RAND's number one customer was the US Air Force. RAND was a researchers' paradise—intellectual freedom, smart colleagues, healthy budgets, and no teaching. Effectively, RAND employees were told:[45]

Figure 5.4 Designers of the Logic Theorist, Allen Newell and Herbert Simon. (*Courtesy Carneige Mellon University.*)

Here's a bag of money, go off and spend it in the best interests of the Air Force.

Simon, the elder of the pair by eleven years, was from Milwaukee. By the 1950s, he was an established political scientist and economist. He was a member of faculty at Carnegie Institute of Technology (CIT) in Pittsburgh and spent summers working at RAND.

Newell grew up in San Francisco, California. He graduated with a degree in Physics from Stanford University before dropping out of an advanced degree in Mathematics at Princeton to join RAND.

The pair first dabbled in computers while working on projects with the goal of enhancing organizational efficiency in air defence centres. RAND's computer, JOHNNIAC, was based on the IAS blueprint. John von Neumann himself was a guest lecturer at RAND. Nonetheless, it was a talk by Oliver Selfridge of MIT Lincoln Labs that captured Newell's imagination. At it, Selfridge described his work on recognizing simple letters (Xs and Os) in images. Newell later reflected:[45]

[The talk] turned my life. I mean that was a point at which I started working on artificial intelligence. Very clear—it all happened one afternoon.

Over the course of the next year, Newell and Simon developed an AI program named Logic Theorist. Newell relocated from Santa Monica to Pittsburgh so as to work more closely with Simon in CIT. Since CIT didn't have a computer, the duo tested their program by gathering a team of students in a classroom and asking them to simulate the behaviour of the machine. The group 'walked' through programs, calling out instructions and data updates along the way. After verification, Simon and Newell transferred the program to Cliff Shaw in RAND Santa Monica via teletype. Shaw entered the program into JOHNNIAC and sent the results back to Pittsburgh for analysis.

The team declared the Logic Theorist operational on 15 December 1955. When the teaching term resumed, Simon was triumphant. He announced: [45]

Over Christmas, Allen Newell and I invented a thinking machine.

Logic Theorist performs algebra on *logic equations*. A logic equation relates variables to one another by means of *operators*. Variables are denoted by letters and can have either true or false values. The most common logical operations are: '=' equals, 'AND', and 'OR'. For example, if we allocate the following meanings to the variables A, B, and W:

$$A = \text{'Today is Saturday'}$$
$$B = \text{'Today is Sunday'}$$
$$W = \text{'Today is the weekend'}$$

we can construct the equation:

$$W = A \text{ XOR } B$$

meaning that 'Today is the weekend' is true if 'Today is Saturday' is true OR 'Today is Sunday' is true, excluding the case that both are true.

By means of algebra, equations such as this can be manipulated so as to reveal new relationships between variables. The series of manipulations that lead from an initial set of equations to a conclusion is called a *proof*. The idea is that if the initial set of equations is valid and the rules of manipulation have been applied properly, the conclusion must also be valid. The starting equations are called the *premises* and the final conclusion the *deduction*. The proof provides formal step-by-step evidence that the deduction is valid given the premise. For example, given:

$$W = A \text{ XOR } B$$

We can prove that:

$$A = W \text{ AND NOT } B$$

In other words, 'Today is Saturday' must be true if 'Today is the weekend' is true and 'Today is Sunday' is false.

Humans produce proofs by intuition and experience. Logic Theorist takes a brute-force approach to finding a proof. It tries all possible algebraic manipulations on the input statements. It repeats this process for the resulting equations and so on. If it finds the conclusion that it is looking for, the search terminates. The program then backtracks and outputs the chain of transformations connecting the deduction to the original premises. This chain is presented to the user as the proof.

Ultimately, Logic Theorist provided step-by-step proofs for thirty-eight of the fifty-two theorems in the classic textbook, *Principia Mathematica*. Indeed, one of Logic Theorist's proofs is more elegant than the textbook version.

In 1959, Newell, Shaw, and Simon introduced a new program. The General Problem Solver used a similar approach to Logic Theorist. However, as the name suggests, the new program tackled a much wider variety of algebraic puzzles, including geometry. To speed up the search, the General Problem Solver algorithm does not try all possible manipulations. It prioritizes equations that are similar to the desired deduction. This means that less time is spent exploring futile paths. Of course, there is an attendant risk that an important line of reasoning is ignored, and the desired conclusion never reached. Rule guided, or *heuristic*, search is now commonplace thanks to its speed.

Newell, Simon, and Shaw's work on reasoning was highly influential. An entire field of AI (*symbolic reasoning*) grew out of the concept of processing logical statements as lists of symbols. Simon even went so far as to claim that the General Problem Solver mimicked human reasoning. Certainly, there are similarities in the manner in which humans sometimes derive formal mathematical proofs by trial-and-error. However, human reasoning appears more intuitive and less rigorous than the General Problem Solver approach.

Newell, the Princeton dropout, was finally awarded a PhD at CIT. In 1967, CIT merged with the Mellon Institute to create the Carnegie Mellon University (CMU). Newell and Simon went on to build one of the world's leading AI research groups at CMU. In 1975, they were awarded the ACM Turing Award for their work on AI and cognitive psychology.

Three years later, Simon received the Nobel Prize for contributions to microeconomics, his other research interest. Newell and Simon lived the rest of their lives in Pittsburgh. Newell passed away in 1992 (aged sixty-five) and Simon in 2001 (aged eighty-four).

Machine Learning

The ability to learn is central to human intelligence. In contrast, early computers could only store and retrieve data. Learning is something entirely different. Learning is the ability to improve behaviour based on experience. A child learns to walk by copying adults and by trial-and-error. Unsteady at first, a toddler's co-ordination and locomotion gradually improve until the infant becomes a proficient walker.

The first computer program to display the ability of learn was unveiled on public television on 24 February 1956. That program was written by Arthur Samuel, of IBM. Samuel's program, like Strachey's, played Checkers. The TV demo was so impressive that it was credited with a fifteen-point uptick in IBM's share price the next day.

Samuel was born in Kansas in 1901. He received a Master's degree in Electronic Engineering from MIT prior to taking up employment with Bell Labs. After the Second World War, he joined the University of Illinois as a Professor. Even though the university lacked a computer, Samuel started work on a Checkers-playing algorithm. Three years later, after joining IBM, Samuel finally got his hands on a real computer. At much the same time as Strachey published his paper on Checkers, Samuel got the first versions of his game-playing program working. On first sight of Strachey's paper, Samuel felt that his own work had been scooped. On closer inspection, it was clear that Strachey's program was a weak Checkers player. Confident that he could do better, Samuel pressed on.

In 1959, Samuel finally published a paper describing his new Checkers program. The understated title—*Some Studies in Machine Learning using the Game of Checkers*—belied the importance of his ideas.

Samuel's algorithm is more thorough in evaluating positions than Strachey's. It achieves this by means of a clever scoring algorithm. Points are given for various on-board *features*. A feature is anything that indicates the strength, or weakness, of a position. One feature is the

Figure 5.5 Originator of machine learning, Arthur Samuel, 1956. (*Courtesy of International Business Machines Corporation,* © *International Business Machines Corporation.*)

difference in the number of checkers that the two players have on the board. Another is the number of crowners. Yet another is the relative positions of checkers. Strategic elements, such as freedom to move or control of the centre of the board, are also considered to be features. Points are scored for every feature. The points for a given feature are multiplied by a *weight*. The resulting values are totalled to give an overall score for the position.

The weights determine the relative importance of each feature. Weights can be positive or negative. A positive weight means that the feature is beneficial for the computer player. A negative weight means that the feature reduces the computer's chances of winning. A large weight means that a feature has a strong influence on the total score. Multiple features with low weights can, however, combine to influence the overall score and thus the final decision.

In summary, Samuel's position evaluation method works as follows:

Take a board position as input.
Set the total score to zero.
Repeat the following for each feature:
    Measure the feature on the board.
    Calculate the points for the feature.
    Multiply by the feature weight.
    Add the result to the total score.
Stop repeating when all features have been scored.
Output the total score.

This scoring mechanism is crucial to Samuel's algorithm. The more accurately the scores reflect the computer's chances of winning, the better the decisions the program makes. Selecting the best features for analysis is important. But beyond that, determining the best weights is essential. However, finding the best values for the weights is tricky.

Samuel designed a *machine learning* algorithm to determine the optimum weights. Initially, the algorithm guesses the weights. The computer then plays a large number of games against itself. One copy of the program plays the white checkers, the other black. As play proceeds, the algorithm adjusts the weights so that the calculated scores more accurately predict the game outcome. If the program wins, the weights that contributed positively to the decision are increased slightly. Similarly, any weights that contributed negatively are decreased. This *reinforces* the winning behaviour. In effect, it encourages the program to play more like this in the future. If the game is lost, the opposite happens. This discourages the program from playing in the same way in the following games. Over a great many games, the learning algorithm fine-tunes the program's play.

The advantages of Samuel's algorithm over manual selection of the weights are twofold. First, the computer never forgets—every single play influences the weights. Second, the computer can play far more games against itself than any human can ever play. Thus, far more information is available to the learning process.

Samuel's development of machine learning was game changing. Previously, altering the behaviour of a program required manual modification of the list of instructions. In contrast, the decisions made by Samuel's program are controlled by the weights, which are simple numerical values. Thus, the behaviour of the program can be adjusted by changing the weights. The program code does not need to be modified. Whereas altering program code is difficult, changing a few numeric weights is trivial. It can be done by an algorithm. This ingenious concept enabled the automation of learning.

In addition, Samuel included a *minimax* procedure for selecting plays. The algorithm performs a lookahead procedure to generate a tree of all possible future moves (Figure 5.6). Scores are calculated for all of the boards at the end of the lookahead. Some of the highest scoring boards are unlikely to occur in a real game since they result from particularly poor play by the opponent. One should assume that both players make good plays. To allow for this, the algorithm backtracks over the tree of plays. The program starts at the leaves of the lookahead tree. On its own turns, the backtracking algorithm selects the play that leads to the maximum score. On its opponent's turns, the program selects the play that leads to the minimum score. At every decision point, the score associated with the chosen play is carried backwards to the parent node. When this backtracking process reaches the root of the tree, the program makes the play associated with the highest backtracked score.

The minimax procedure functions as follows:

Take the tree of possible plays as input.
Start at the penultimate layer.
Repeat the following for every layer:
    Repeat the following for every node in the layer:
        If it is the computer's turn,
        then select the play giving the maximum score,
        else select the play giving the minimum score.
        Copy the minimax score to the current node.
    Stop repeating when all nodes in this layer have been
        examined.
Stop repeating when the root of the tree is reached.
Output the play with the maximum backtracked score.

Image a simple two-ply lookahead (Figure 5.6). The tree includes the computer's potential next plays and its opponent's possible responses. The scores at the leaves of the tree (ply 2) are inspected to find the minimum in each sub-tree. This reflects the opponent's selection of the best play from their point of view. These minimum scores are copied

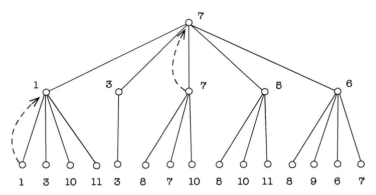

Figure 5.6 Lookahead tree showing backtracking scores obtained using the minimax algorithm.

to the nodes immediately above (ply 1). This puts the scores 1, 3, 7, 5, and 6 on the nodes in ply 1. Now, the algorithm selects the play giving the highest score. This means that the computer chooses the best play from its point of view. Thus, the maximum value of 7 is copied back to the root of the tree. The play leading to the board with the score of 7 is the best choice, provided that the opponent is a good player. This play forces the opponent into a choice between positions with scores of 8, 7, and 10. The best that the opponent can do is accept the position with a score of 7.

To make effective use of the available compute time, Samuel's program adjusts the depth and width of the lookahead search according to a set of rules (i.e. it uses heuristic search). When a position is unstable, for example just before a jump, the program looks further ahead. Bad plays are not explored in depth. *Pruning* the search in this way affords more time for evaluation of likely scenarios. To further accelerate processing, Samuel's program stores the minimax scores for commonly occurring board positions. These scores do not need to be recalculated during execution, as simple table look-up suffices.

In 1962, Samuel's Checkers playing program was pitted against Robert Nealey, a blind Checkers master. The computer's victory was widely hailed but Nealey wasn't even a state champion. It would be thirty years (1994) before a computer program finally defeated the Checkers world champion.

Samuel retired from IBM in 1966 to take up a research professorship at Stanford University. He was still programming at the age of eighty-five when Parkinson's disease finally forced him to stop working. Samuel passed away in 1990.

The rudiments of today's most advanced board game playing algorithms can be seen in Samuel's 1950s work. Minimax, reinforcement learning, and self-play are the basis of almost all modern Checkers, Chess, and Go playing AIs. Moreover, as we shall see, machine learning has proven to be incredibly effective in tackling complicated data analysis problems in a great many applications.

The AI Winters

During the late 1950s and 1960s, expectations for AI were sky high. Supported by Cold War military funding, AI research groups flourished, most notably at MIT, CMU, Stanford University, and Edinburgh University. In 1958, Newell and Simon predicted that within just ten years:[83]

> [A] digital computer will be the world's Chess champion, unless the rules bar it from competition.

Four years later, Claude Shannon, founder of information theory, pronounced deadpan to a television camera:[84]

> I confidently expect that within a matter of ten or fifteen years something will emerge from the laboratory which is not too far from the robot of science fiction fame.

In 1968, Marvin Minsky, Head of AI Research at MIT, predicted that:[85]

> [I]n thirty years we should have machines whose intelligence is comparable to man's.

Of course, none of these predictions came true.

Why were so many eminent thinkers so spectacularly wrong? The simplest answer is hubris. These were mathematicians. To them, mathematics was the pinnacle of intelligence. If computers could perform arithmetic, algebra, and logic, then surely more mundane forms of intelligence must soon yield. What they failed to appreciate was the variability of the real-world and the complexity of the human brain.

Processing images, sounds, and language turned out to be much more complicated than dealing with equations. The endemic failure of AI projects caused the funding agencies and politicians to question the value of this type of research. In the UK, the British Science Research Council asked Sir James Lighthill, Lucasian Professor of Mathematics at Cambridge University, to lead a review of AI research. Published in 1973, his report was damning:[86]

> In no part of the field have the discoveries made so far produced the major impact that was then [(around 1960)] promised.

Funding for AI research in the UK was savagely cut. Meanwhile, in the shadow of the Vietnam War, the Mansfield Amendments of 1969 and 1973 curtailed US governmental spending on research. Only projects with direct military applications were to be given money.

The first of a succession of 'AI Winters' set in. Starved of resources, AI research groups shrank and shrivelled.

With AI in the doldrums, computer science pivoted to more practical applications. Given the limitations of computer performance, some scientists sought to develop fast algorithms to solve important, but computationally complex, problems. This quest for speed was to lead to one of the greatest unsolved enigmas in mathematics.

6

Needles in Haystacks

Der Handlungsreisende – wie er sein soll und was er zu thun hat,
um Aufträge zu erhalten und eines glücklichen Erfolgs in seinen
Geschäten gewiß zu sein.

EIN ALTER COMMIS-VOYAGEUR
Mit einem Titelkupfer, 1832 [87]

In the 1970s, one of the greatest mysteries in mathematics was uncovered by an assemblage of researchers investigating the properties of
algorithms. Despite a $1 million prize, that mystery remains unsolved.
At the very heart of the matter is a seemingly innocuous problem.

The Travelling Salesman Problem

The Travelling Salesman Problem asks that the shortest tour of a set of
cities is determined. The cities' names and the distances between them
are provided. All the cities must be visited once, and only once. The cities
can be visited in any order, as long at the trip starts from, and ends at,
the salesman's home city. The challenge is to find the tour that offers
the shortest total distance travelled.

The Travelling Salesman Problem was first documented in the 1800s.
At the time, it was a practical concern for commercial travellers jour-
neying between the cities of continental Europe. Later, the Problem was
reformulated as a mathematical plaything by William Hamilton and
Thomas Kirkman.

Let's say that Berlin is the titular salesman's home and he must visit
Hamburg, Frankfurt, and Munich (Figure 6.1). The simplest way to find
the shortest tour is by means of an *exhaustive* search. Exhaustive, or brute-
force, search calculates the length of every possible tour and picks the
shortest. An exhaustive search algorithm proceeds as follows:

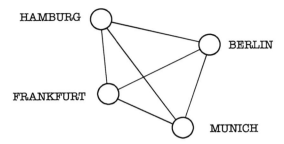

Figure 6.1 The Travelling Salesman Problem: Find the shortest tour that visits every city once and returns home.

> Take a set of city names as input.
> If there is only one city in the set,
> then output a tour containing that city alone,
> else:
>> Create an empty list.
>> Repeat the following for every city in the set:
>>> Create a copy of the set, omitting the selected city.
>>> Apply this algorithm to the reduced set.
>>> Insert the selected city at the start of all the tours
>>>   returned.
>>> Append these tours to the list.
>> Output all tours found.

To start with, the set of all the cities, excluding the home city, is input to the algorithm. The home city is the known start and end point of every tour, so it does not need to be included in the search. The algorithm creates a tree of city visits from the input set (Figure 6.2). The algorithm relies on two mechanisms. First, it uses repetition—the algorithm selects every city in the input set, one after another, as the next to be visited. Second, it uses recursion (see Chapter 1). For each city, the algorithm calls a copy, or *instance*, of itself. An instance of an algorithm is another, separate enactment of the algorithm that operates on its own data. In this case, every instance creates a new sub-tree in the diagram. After each city is visited, the set of cities input to the next instance of the algorithm is reduced. Thus, the instances deal with fewer and fewer cities until there is just one left in the set. When this happens,

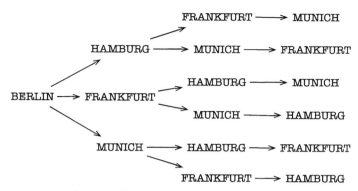

Figure 6.2 Tree showing all possible tours. All tours end in Berlin (not shown).

the tree's leaf instance terminates, returning a tour containing just one city. The previous instances of the algorithm take this output and add the selected cities in reverse order. In this way, the algorithm unwinds, creating tours as it moves up the tree. Once all of the tours have been traced back to the root of the tree, the original instance of the algorithm terminates, and the completed list of tours is output.

As the list of tours is being generated, the length of the tours is calculated by totalling the city-to-city distances.

The operation of the algorithm can be visualized as an animation. The algorithm constructs the tree from the root. From there, it grows the topmost path, one city after another, until the top leaf is reached. It then backtracks one layer and grows the second leaf. Next, it goes back two layers, before adding the third and fourth leaves. The algorithm continues sweeping to and fro until the entire tree has been created. In the end, the algorithm returns to root and terminates.

In the example, Berlin is selected as the home city and so is excluded from the input set of {Hamburg, Frankfurt, Munich}. The first instance of the algorithm selects Hamburg, Frankfurt, and Munich in turn as the first city. For each of these selections, the algorithm spawns a new instance to explore a sub-tree. After selecting Hamburg as the first city, the second instance of the algorithm chooses Frankfurt from the set {Frankfurt, Munich}. It then creates an third instance to deal with the remaining city: {Munich}. Since there is only one city:

Munich

is returned as the only possible tour. The calling instance then prepends Frankfurt, producing the tour:

Frankfurt, Munich.

The same instance then explores the alternative branch, giving the tour:

Munich, Frankfurt.

These partial tours are returned to the calling instance which adds its selection, giving the tours:

Hamburg, Frankfurt, Munich;
Hamburg, Munich, Frankfurt.

The sub-trees starting with Frankfurt and Munich are explored in a similar way. Finally, the complete list of tours is output and the original algorithm instance terminated.

An exhaustive search such as this is guaranteed to find the shortest tour. Unfortunately, brute-force is slow. To figure out just how slow it is, we need to think about the size of the tree. In the example, the roadmap contains four cities. The cities are fully connected in that every city is directly connected to all others. Thus, on leaving Berlin, there are three possible stops. For each of these stops, there are two possible next cities since the salesman can't go back home yet or return to the start. After the first and second stops, there is only one possible third city. Expanding this out, gives the number of possible tours as $3 \times 2 \times 1 = 6$, that is, 3 factorial (3!).

Computing the lengths of six tours is a manageable manual computation. What happens if there are 100 cities? One hundred fully connected cities would give ninety-nine factorial tours, approximately 9×10^{155} (a 9 with 155 zeroes after it). A modern desktop computer couldn't possibly cope with that! For the Travelling Salesman Problem, exhaustive search is surprisingly slow even for roadmaps of seemingly moderate size.

While more efficient algorithms have been found, the quickest isn't much faster than exhaustive search. The only way to significantly speed up the search is to accept a compromise. You have to accept that the algorithm might not find the shortest possible tour. To date, the best fast approximation algorithm is only guaranteed to find a path within forty per cent of the minimum. Of course, compromises and approximations are not always acceptable. Sometimes, the shortest possible path must be found.

Over the years, researchers have experimented with programs that hunt for the shortest tours of real roadmaps. At the beginning of the computer age (1954) the largest Travelling Salesman Problem with a known solution contained just forty-nine cities. Fifty years later, the largest solved tour contained 24,978 Swedish cities. The current state-of-the-art challenge is a world map of 1,904,711 cities. The shortest tour found on that map traverses 7,515,772,212 km. Identified in 2013 by Keld Helsgaun, no one knows if it is the shortest possible tour or not.

Measuring Complexity

The trouble with the Travelling Salesman Problem is the computational complexity of the algorithm needed to solve it. Computational complexity is the number of basic operations—memory accesses, additions, or multiplications—required to perform an algorithm. The more operations an algorithm requires, the longer it takes to compute. The most telling aspect is how the number of operations grows as the number of elements in the input increases (Figure 6.3).

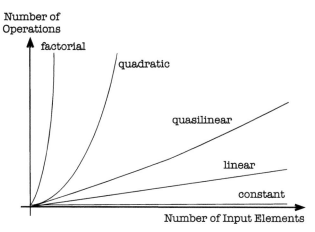

Figure 6.3 Graph showing relationships between computational complexity and the number of inputs to an algorithm.

A straightforward algorithm has *constant* complexity. For example, adding a book to the top of an unsorted pile of novels has a computational complexity of just one operation. The complexity of adding the book to the pile is fixed, regardless of the number of books already in the stack.

Finding a particular title in a bookcase full of books takes more operations. If the books are unsorted, the librarian might have to check every single title before the sought-after tome is found. Putting an additional book in the bookcase increases the worst-case computational complexity of the search by one operation. In other words, the complexity of title search grows in proportion to the number of books on the shelves. For this problem, computational complexity is *linear* with the number of books.

Algorithms for sorting books have still greater complexity. Insertion Sort (see Introduction) organizes books one at a time. Putting a book on the shelf requires that all books already in place are either scanned past or shifted over. As a consequence, Insertion Sort has computational complexity proportional to the number of books squared. This gives a *quadratic* relationship between the number of books and the number of operations.

As might be expected, Quicksort (see Introduction) has lower complexity. Quicksort repeatedly splits books into piles based on selected pivot letters. When the piles contain five or fewer books, Insertion Sort is applied to every pile and the piles are transferred in order to the shelf. On average, the complexity of Quicksort is equal to the number of books multiplied by the logarithm of the number of books. Since the logarithm of a quantity grows more slowly than the quantity itself, Quicksort's complexity is lower than that of Insertion Sort. Quicksort's average computational complexity is *quasilinear*.

Algorithms for adding, searching for, and sorting books have what is called *polynomial* computational complexity. A polynomial time algorithm has computational complexity proportional to the number of inputs to some constant power. In the constant complexity case, the power is zero. For linear, it is one and for quadratic, the power is two. Polynomial time algorithms can be slow for very large numbers of inputs, but in the main, they are tractable on modern computers.

The more challenging algorithms are those with *superpolynomial* computational complexity. These methods have complexity in excess of polynomial time. The number of operations needed to perform superpolynomial time algorithms explodes as the number of inputs increases. The exhaustive search algorithm for solving the Travelling Salesman Problem has superpolynomial time complexity. As we have seen, the number of operations needed is equal to the number of cities factorial. Adding a single city to a tour multiplies the number of operations required by the number of cities already on the map. This multiplicative effect causes an extremely rapid expansion in complexity as the number of cities grows.

Much work has been conducted on reducing the computational complexity of such algorithms. The tricks that can be played depend on the specifics of the problem. Sometimes the structure of the input data can be exploited to quickly identify obvious solutions or partial solutions. In other cases, using additional data storage enables reductions in the number of operations. In much the same way, the index of a book increases the number of pages but greatly reduces the time needed to find a given keyword.

It is clear that for every problem, there must be a fastest algorithm. The trick is to find it. In the 1960s and 1970s, a handful of theoreticians began to investigate the limits of algorithmic complexity. Much of what we now know about the topic stems from that formative work.

Complexity Classes

Computational problems are graded according to the complexity of the fastest known algorithms that solve them (Figure 6.4; Table 6.1). Problems that can be solved with polynomial time algorithms are referred to as *P problems* (polynomial time). P problems are considered quick to solve. For example, sorting is a P problem.

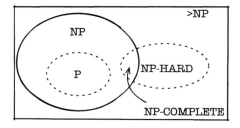

Figure 6.4 Common problem complexity classes.

Table 6.1 Table listing the common complexity classes.

Class	Solution Time	Verification Time
P	polynomial	polynomial
NP	not specified	polynomial
NP\P	> polynomial	polynomial
>NP	> polynomial	> polynomial
NP-Complete	most complex P	polynomial
NP-Hard	P convert to NP-Complete	not specified

Problems whose solutions can be verified using a polynomial time algorithm are referred to as *NP problems* (non-deterministic polynomial time). How long NP problems take to solve is not defined. Some can be solved in P time; others cannot. Since solving a problem is a way to verify a given solution, P problems are by definition also NP problems. In other words, the set of all P problems is a subset of the set of NP problems.

Problems that are in NP but not in P are called *NP\P problems* (NP minus P). These problems are slow to solve but the answers, when they are known, are quick to check. Sudoku is an NP\P problem.

Sudoku is a Japanese number puzzle played on a 9x9 grid. At the start of the puzzle, some squares are blank, and some contain digits. The goal is to fill every blank with a digit in the range 1 to 9. The difficulty is that a digit may only appear once per row and once per column. A brute-force approach tries all possible digit placements. As a consequence, brute-force search is slow. Its complexity is super-polynomial with grid size. In contrast, verifying a completed grid is quick. The checker need only scan the rows and columns for blank squares and duplicated digits. Thus, checking can be completed in polynomial time. Like other NP\P problems, Sudoku is slow to solve but quick to check.

Beyond this, are *greater than NP problems* (>NP). These are problems that cannot be solved or verified in polynomial time. Their solvers and checkers require superpolynomial time. The Travelling Salesman Problem is one such problem: solving it requires factorial time. The only way to check that an answer is the minimum tour is to run the solver again. Thus, verifying the solution also takes factorial time. Slow to solve and slow to verify problems are >NP.

In 1971, Stephen Cook of the University of Toronto, Canada published a paper that was to have major ramifications for complexity. Cook had just joined the University, having been passed over for tenure at the University of California, Berkeley.[91] Cook's paper uncovered deep relationships between certain problem types. The paper led to one of the greatest mysteries in mathematics, the so-called 'P versus NP Problem', which asks, 'Can a polynomial time algorithm be found that will solve all NP problems?'

The importance of the question is clear. A polynomial time algorithm for solving all NP problems would revolutionize a host of applications. Previously intractable problems could be quickly solved. For example, more efficient schedules could be derived in industries ranging from transportation to manufacturing. Molecular interactions could be predicted, accelerating drug design and solar panel development.

The most complex NP problems are called the *NP-Complete problems*. Cook showed that a polynomial time algorithm for solving an NP-Complete problem could be used to solve all NP problems in polynomial time. In other words, a fast algorithm for solving one NP-Complete problem would, by extension, turn all NP\P problems into mere P problems. As a result, the set NP would suddenly equal the set P.

Furthermore, Cook's work proved that some >NP problems can be *transformed*, in polynomial time, into NP-Complete problems. This transformation is achieved by processing the >NP problem's inputs in such a way that an NP-Complete algorithm can finish the computation. Therefore, finding a fast algorithm for solving an NP-Complete problem would provide faster solvers for these >NP problems. Collectively, this group of NP-Complete and >NP problems that are amenable to P-time transformation to NP-Complete problems are called the *NP-Hard problems*. A polynomial time algorithm for solving an NP-Complete problem would lead to fast solvers for all NP-Hard problems, as well.

The NP-Complete problems are revered. A polynomial time algorithm for solving any one of them would likely win a Fields Medal, the mathematical equivalent of the Nobel Prize.

The Travelling Salesman Problem is NP-Hard. A simplified version of the Travelling Salesman Problem, the Travelling Salesman Decision Problem, is known to be NP-Complete. The problem asks, 'For a given roadmap, can a tour be found which is less than a specified target distance?' Whereas, verifying a solution to the Travelling Salesman

Problem requires solving the problem again (>NP), verifying a given solution to the decision problem is fast—simply measure the length of the given tour and compare the result to the specified distance. Thus, the Decision Problem is slow to solve but fast to verify (NP\P). The set of NP-Complete problems includes the Knapsack Packing Problem, the game of Battleships, and the Graph Colouring Problem.

In the year 2000, the Clay Mathematics Institute of Cambridge, Massachusetts announced $1 million prizes for solutions to seven Millennium Problems. These Problems were selected as being the most important in all of mathematics. The 'P versus NP Problem' was one of the seven. The Institute offers the prize to anyone who can provide a polynomial time algorithm for solving an NP-Complete problem, or a definitive proof that no such algorithm exists.

Most researchers now think that P is not equal to NP, and never will be. This conclusion stems from nigh on forty years of failed attempts to find fast algorithms to solve NP-Complete problems. On the other hand, a proof that such an algorithm cannot exist remains elusive. So far, the $1 million Clay Mathematics Institute prize remains unclaimed.

Cook received the ACM Turing Award in 1982. In the citation, he was lauded for transforming 'our understanding of the complexity of computation'. Afterwards, Richard Kamp, Professor Emeritus of Electronic Engineering and Computer Science at the University of California, Berkeley, ruefully wrote: [91]

> It is to our everlasting shame that we were unable to persuade the Math department to give him tenure.

Short Cuts

The Travelling Salesman Problem is one of many *combinatorial optimization* problems, which require that a number of fixed elements be combined in the best way possible. In the case of the Travelling Salesman Problem, the fixed elements are the city-to-city distances and 'the best way possible' is the shortest tour. The fixed elements can be arranged in a myriad of ways. The goal is to find the single, best arrangement.

Practical combinatorial optimization problems abound. How best to allocate staff to tasks in a large factory? What flight schedule maximizes revenue for an airline? Which taxi should pick up the next customer

Figure 6.5 Edsger Dijkstra, inventor of the route-finding algorithm, 2002. (© *2002 Hamilton Richards.*)

so as to maximize profits? All of these questions require a fixed set of resources to be assigned in the best way possible.

As we have seen, brute-force search algorithms try out all possible combinations and select the best. In 1952, Edsger Dijkstra (Figure 6.5) came up with a fast algorithm that solves the world's most common combinatorial optimization problem. His algorithm is now embedded in billions of electronic devices.

Dijkstra was Holland's first professional programmer. Having completed a programming course in England in 1951, he was offered a part-time job as a computer programmer in the Mathematisch Centrum (Mathematical Centre) in Amsterdam. Inconveniently, the Centrum didn't have any computers. Short on funding, and these being the early days of computing, the Centrum was in the process of building one. Dijkstra worked part-time at the Centrum while studying Mathematics and Physics at the University of Leiden. After three years, he felt he couldn't keep both programming and physics going. He had to choose one or the other. He loved programming, but was it a respectable profession for a serious young scientist? Dijkstra went to see Adriaan van Wijngaarden, the Director of the Computation Department. Van Wijngaarden agreed that programming was not a discipline in its own right, just yet. However, van Wijngaarden confidently predicted that computers were here to stay and that this was merely the beginning. Could Dijkstra not be one of those who turned programming into a respectable discipline? By the time he left van Wijngaarden's office an hour later, Dijkstra's path in life was set. He completed his studies in physics as quickly as he could.

One year on and Dijkstra was in another predicament. The Centrum was due to host some important guests. The highlight of their visit was to be the Centrum's now operational computer. Dijkstra was asked to give a demonstration of the machine's capabilities. Since his guests knew little about computers, Dijkstra decided that his demo should focus on a practical application. He hit upon the idea of writing a program to determine the shortest driving route between two Dutch cities. While he was satisfied with the concept, a difficulty remained. There was no fast algorithm for finding the shortest route between two cities.

While out shopping one morning, he and his fiancée stopped at a cafe. Whilst sipping coffee on the terrace, Dijkstra invented an efficient route-finding algorithm in about twenty minutes. It would be three years before he got around to publishing it. It just didn't seem particularly important.

Dijkstra's algorithm is akin to playing a board game. It finds the shortest route by moving a token between cities on a roadmap. As the token moves across the map, the cities are annotated with the route taken and the accumulated distance from the starting point. When the token leaves a city, the city's name is crossed off a list of cities to visit so that the token cannot return.

To begin, the token is placed on the city of origin. The city's name and zero distance are recorded next to it. Every directly connected city is then considered. The distance travelled from the starting point to these cities is computed. This is achieved by adding the number beside the token to the length of the link from the token to the directly connected city. If the city is already annotated with a distance less than this computed value, the existing annotation is left unchanged. If the newly calculated value is less than the recorded distance, the annotation is replaced. The new distance is written down along with the route taken. The route taken is the list of cities beside the token, followed by the name of the new city. When these steps have been carried out for all directly connected cities, the token is forwarded to the city that has the least annotated distance and that has not yet been visited. This process—check and update the directly connected cities, shift the token—is repeated until the token arrives at the desired destination.

Imagine Dijkstra's algorithm operating on a Dutch roadmap (Figure 6.6). Let's say that the point of departure is Amsterdam and the

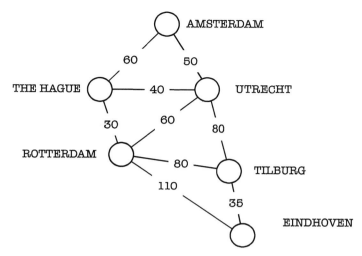

Figure 6.6 The Routing Finding problem requires finding the shortest route between two cities. This roadmap shows the distances between the principal Dutch cities in kilometres.

destination is Eindhoven. To start, the token is placed on Amsterdam. The capital is annotated with:

Amsterdam 0.

The Hague and Utrecht are directly connected to Amsterdam and so are annotated with:

Amsterdam-The Hague 60;
Amsterdam-Utrecht 50.

The total distance to Utrecht is least, so the token is moved there. The cities directly connected to Utrecht are then considered. Thus, Tilburg and Rotterdam are annotated with:

Amsterdam-Utrecht-Tilburg 50+80=130,
Amsterdam-Utrecht-Rotterdam 50+60=110.

The Hague and Amsterdam are not updated since the distances to them from Amsterdam via Utrecht (90 and 100, respectively) are greater than the paths already recorded (60 and 0).

The token is transferred to The Hague, the city with the shortest cumulative distance (60) which has not yet been visited. Going from

Amsterdam to Rotterdam via The Hague is a shorter journey than the previous trip through Utrecht. Hence, Rotterdam is updated with:

Amsterdam-The Hague-Rotterdam 60+30=90.

The token is now moved to Rotterdam. Directly connected Eindhoven is then updated with:

Amsterdam-The Hague-Rotterdam-Eindhoven 90+110=200.

The path to Tilburg is longer (170 via Rotterdam) than the annotation already in place (130 via Utrecht), so Tilburg is not updated.

The as-yet unvisited city with the least accumulated distance is Tilburg, so the token is shifted there. The total distance to Eindhoven via Tilburg is 165, which is less than the current annotation (200 via Rotterdam). As a result, Eindhoven's note is replaced with:

Amsterdam-Utrecht-Tilburg-Eindhown 165.

The token is transferred to Eindhoven and the algorithm is complete. The shortest path is Amsterdam-Utrecht-Tilburg-Eindhoven with a total distance of 165 km.

Substituting time for distance allows Dijkstra's algorithm to find the fastest route instead of the shortest route.

Dijkstra algorithm is important for two reasons. First, it is guaranteed to find the shortest route. Second, the algorithm is fast. It prunes the search on the fly, avoiding bad solutions and concentrating its efforts on good ones. Third, routing problems are ubiquitous. Every person and vehicle in the world must navigate.

Dijkstra's algorithm soon became popular in computing circles. In 1968, it was enhanced by three researchers working at the Stanford Research Institute (SRI) in California. The researchers were part of the team that built Shakey the Robot. Shakey was the first general-purpose mobile robot with reasoning capability. Clunky by today's standards, Shakey was essentially a large box containing a small computer mounted on powered wheels. Its pointy metal 'head' supported a large video camera and an ultrasonic range finder. Preposterous as it now seems, *Life* magazine went so far as to call Shakey the 'first electronic person'.

Since Shakey was mobile, it needed to be able to navigate. While incorporating this functionality, the development team spotted an inefficiency in Dijkstra's algorithm. The algorithm occasionally wastes time moving the token to cities that lead away from the final destination. These cities look promising as they have short links to the city marked with the token. However, they lead in the wrong direction and are ultimately eliminated. To remedy this flaw, Peter Hart, Nils Nilsson, and Bertram Raphael proposed the A* (*a-star*) algorithm. A* utilizes a modified distance metric. In Dijkstra's original algorithm, the metric is the distance travelled. In A*, the metric is the distance travelled so far plus the straight-line distance from the current city to the final destination. Whereas Dijkstra's algorithm only considers the path so far, A* estimates the length of the complete route, from start to finish. As a result, A* is less inclined to visit cities that take the token away from the destination.

Nowadays, variants of A* are used in every navigation app on the planet, from sat navs to smartphones. To allow greater accuracy, cities have since been replaced by road intersections, but the principles remain the same. As we will see, derivatives of Dijkstra's algorithm are now used to route data over the Internet.

In fulfilment of van Wijngaarden's prophecy, Dijkstra went on to make a series of significant contributions to the precarious emerging discipline that became computer science. Most notably, he originated algorithms for *distributed computing*, whereby multiple computers cooperate to solve highly computationally complex problems. In recognition of his work, Dijkstra was the 1972 recipient of the ACM Turing Award.

Stable Marriages

Roadmaps are not the only source of combinatorial optimization problems. Matching problems seek to pair items in the best way possible. A classic matching problem is pairing hopeful college applicants with available places. The challenge is to assign high school graduates to courses in a way that is both fair and satisfies as many students and colleges as possible. As with other combinatorial optimization problems, matching becomes difficult as the number of items increases. Fast algorithms are mandatory even for moderate numbers of students.

The seminal paper on matching was published by David Gale and Lloyd Shapley in 1962. The pair struck up a friendship at Princeton University, New Jersey, where both studied for PhDs in Mathematics. After Princeton, Gale joined Brown University in Rhode Island,

and Shapley moved to the RAND Corporation. Gale and Shapley's paper tackled the age-old problem of matching dating singles for marriage.

The Stable Marriage Problem seeks to pair men and woman for marriage. At the outset, female participants rank all males according to their own preferences (1st, 2nd, 3rd, and so on), and, vice versa, the men rank the women. The objective is to pair the participants in such a way that the set of marriages is stable. The set is considered stable if there exists no male and female who prefer each other to their spouses. Such a circumstance could lead to divorce.

Gale and Shapley's paper proposed a remarkably simple algorithm for solving the Stable Marriage Problem. So simple, in fact, that the authors had trouble getting it published at all. In the paper, Gale and Shapley used the Stable Marriage Problem as a proxy for a range of real-world *two-way* matching problems. Two-way refers to the fact that both parties state preferences, not just one party. The Gale–Shapley algorithm quickly became the de facto two-way matching method. The algorithm is still employed in a wide variety of applications, including matching critically ill patients with organ donors.

The Gale–Shapley algorithm matches men and women over successive rounds. In each round, all of the unattached men make one marriage proposal. (For simplicity, the paper assumes that the participants are heterosexual and that the marriage proposals are made by the men. In fact, the proposals could equally be made by the women—it doesn't matter to the algorithm.) Every unattached man proposes to the woman who has his highest preference and has not previously rejected him. If the woman receiving the proposal is not engaged, she automatically accepts the proposal. If she is already engaged, she considers her preferences for her new suitor and her existing fiancée. If she prefers her new suitor, she spurns her fiancée and is betrothed to her new paramour. If she prefers her existing fiancée, she turns down the advances of the would-be interloper. Once her decision is made, the algorithm continues to the next unmarried man. When all of the unmarried men have made proposals, the algorithm moves on to the next round. At this stage, a rebuffed male fiancée is free to make a fresh proposal of marriage to someone else. The algorithm ends when all participants are engaged for marriage.

Consider six friends—three men and three woman—living across the hall from each another in two New York apartments close to

Table 6.2 Table showing marriage preferences.

Preference	Alex	Ben	Carlos	Diana	Emma	Fiona
1st	Diana	Diana	Diana	Ben	Ben	Carlos
2nd	Emma	Fiona	Fiona	Alex	Alex	Alex
3rd	Fiona	Emma	Emma	Carlos	Carlos	Ben

Central Park. For the sake of discretion, let's call them Alex, Ben, Carlos, Diana, Emma, and Fiona. All know one another other. All are single. All have matrimony on their minds. When asked, they state the marriage preferences listed in Table 6.2.

In the first round, Alex proposes to Diana. Diana accepts because she is currently available. Next, Ben proposes to the popular Diana. Since Diana prefers Ben to Alex, she ditches the latter and accepts Ben's proposal. Carlos also proposes to Diana and gets turned down flat. In the second round, Alex and Carlos are unattached. Alex proposes to Emma, number two on his list. Emma is, as yet, without a partner, so she acquiesces to Alex's advances. Carlos requests Fiona's hand in marriage and she agrees since she is unattached. That's it. Everyone is now engaged—Alex & Emma, Ben & Diana, Carlos & Fiona. All of the marriages are stable. Emma would prefer Ben but she can't have him since he would rather couple with Diana, his soon-to-be wife. Similarly, Alex and Carlos have unrequited feelings for Diana but she is set to marry the man of her dreams, Ben. Fiona is happily betrothed to Carlos—her number-one pick—even though he, also, has a crush on Diana.

As a dating scheme, the Gale–Shapley algorithm is brutal—all those rejections! Nevertheless, one wonders if humans seeking partners intuitively follow a procedure akin to the Gale–Shapley algorithm. In real life, explicit proposals and firm replies become surreptitious smiles, longing gazes, inquiries via friends, and polite refusals. While there are similarities, there are differences. In truth, preferences evolve over time. The emotional cost of separation means that individuals are reluctant to make drastic changes. Despite these contrasts, some online dating agencies now use the Gale–Shapley algorithm to match clients.

One of the biggest matching exercises in the world is carried out every year by the National Residency Matching Program (NRMP). The program pairs medical school graduates with internship opportunities in hospitals across the US. Currently, the program matches 42,000 graduate applicants with 30,000 hospital vacancies per annum. When it was established in 1952, the NRMP adopted a matching algorithm from a previous clearing house. The Boston Pool algorithm was used by the NRMP for forty years. During the 1970s, it came to light that the Boston Pool algorithm was, in fact, the same as the independently developed Gale–Shapley method. Embarrassingly, the unknown Boston Pool team had beaten the eminent mathematician–economist duo to the punch by more than a decade. Of course, their academic paper included a formal proof, whereas the Boston Pool algorithm was ad hoc.

In the 1990s, the noted economist and mathematician Alvin Roth was engaged to revamp the NRMP matching algorithm. Moving with the times, Roth's new method allows medical couples to seek co-location. It also seeks to prevent rogue applicants gaming the system to their advantage. Unlike the Boston Pool, Roth's technique relies on one-way matching, whereby only the applicants' preferences are considered.

Shapley and Roth were jointly awarded the Nobel Prize in Economics in 2012 for their work on *game theory*. This is the branch of mathematics concerned with competition and cooperation between intelligent decision makers. Shapley, the elder of the two, is widely seen as the grand theoretician who laid the groundwork for Roth's practical studies on how markets operate. One of the highlights of their Nobel Prize citation was the Gale–Shapley algorithm. Gale passed away in 2008 and so was ineligible for the Prize. Shapley died in 2016, at the grand old age of 92. Roth continues to work at Stanford and Harvard Universities.

Artificial Evolution

In the 1960s, John Holland (Figure 6.7) took a radical approach to solving combinatorial optimization problems. Uniquely, his algorithm was four billion years old!

Holland was born in Fort Wayne, Indiana in 1929. Like Dijkstra, Holland caught the computer programming bug while studying physics. At MIT, he wrote a program for the Whirlwind computer. Funded by the US Navy and Air Force, Whirlwind was the first real-time computer to incorporate an on-screen display. The machine was designed to process radar data and provide early warning of incoming aircraft and missiles. After a brief sojourn programming at IBM, Holland moved to

the University of Michigan to pursue a Master's degree followed by a
PhD in Communication Sciences. It would be several years before the
term 'computer science' came into vogue. Holland's PhD supervisor
was Arthur Burks, the man who ran the ENIAC demo for the press in
1946.

While browsing in the university library, Holland came upon an old
book by Ronald Fisher called *The Genetical Theory of Natural Selection* (1930).
The book employed mathematics to investigate natural evolution. Holland later recollected: [103]

> That was the first time I realized that you could do significant mathematics on evolution. The idea appealed to me tremendously.

Inspired, Holland determined that he would replicate evolution in
a computer. He pursued this singular idea throughout the 1960s and
1970s. Convinced he was on to something significant, Holland authored
a book detailing his findings in 1975. Sales were disappointing. The
research community wasn't much interested.

Almost twenty years later, Holland penned an article on the topic
of *genetic algorithms* for the popular science magazine *Scientific American*.
A second edition of his book went to press that same year. Finally,
genetic algorithms broke through to the mainstream of computing
research. Holland's long-neglected text has now been cited (formally
referenced) in more than 60,000 books and scientific papers—a box
office hit by academic standards.

Figure 6.7 Designer of the first genetic algorithms, John Holland. (© *Santa Fe Institute.*)

Natural evolution adapts a species to its environment by means of three mechanisms—*selection, inheritance,* and *mutation.* The process acts on a species living in the wild. While the individuals in a species share many characteristics, there is variation from one animal to the next. Some traits are beneficial for survival, while others are detrimental. Selection refers to the fact that individuals with beneficial characteristics are more likely to survive to adulthood and reproduce. Inheritance is the tendency of children to display similar physical traits as their parents. Thus, children of survivors tend to have the same beneficial characteristics. Mutation is the manifestation of random alterations in the genetic material passed on to a child. Depending on what part of the chromosome is affected, a mutation might have no effect, a limited impact, or cause an extreme change. Over many generations, selection and inheritance mean that a population will tend to have a greater proportion of individuals with traits that are beneficial to survival and reproduction. Mutation is the joker in the pack. Mostly, it has no impact on the population. Sometimes, it plants the seed for a radical beneficial change.

The classic example of natural evolution at work is the pepper moth. The name comes from the insect's mottled wings which look like black pepper sprinkled on white paper. The insect's appearance is a form of camouflage, which makes it difficult for predatory birds to spot the creatures against the bark of the local trees. In eighteenth-century England, peppered moths were predominantly pale in colour. Strangely, by the end of the nineteenth century, almost all peppered moths in the major cities were dark. Over the course of a century, the English urban pepper moth population had changed colour.

It became apparent that the change was precipitated by the Industrial Revolution. Rapid exploitation of fossil fuels had led to the construction of large numbers of factories belching out great plumes of smoke and soot. Tree bark, walls, and lampposts gradually turned black. Pale moths became more vulnerable to predators and were killed in greater numbers. The dark moths thrived, passing their natural camouflage on to their children. Over time, the balance in numbers within the population shifted in favour of the dark moths.

Holland believed that artificial evolution could be used to solve combinatorial optimization problems. His idea was that possible solutions could be thought of as individuals within a population. Drawing on genetics, he suggested that every solution be represented by a sequence of letters. For example, to solve the Travelling Salesman Problem, a tour might be represented using the first letter of the cities: BFHM. He considered this sequence analogous to a living organism's chromosome (DNA).

Holland's genetic algorithms act on this pool of *artificial chromosomes*. Selection is performed by evaluating every artificial chromosome and discarding the worst performing ones. Inheritance is mimicked by intermingling letter sequences to create the next generation of chromosomes. Mutation is replicated by randomly replacing letters in a handful of chromosomes. These three processes are repeated to produce generations of chromosomes until, finally, an acceptable solution is identified among the population.

Over many generations, the three mechanisms act in concert to increase the proportion of good solutions in the population. Selection guides the search towards better solutions. Inheritance mixes promising answers in unexpected ways to produce new candidate solutions. Mutation raises the diversity in the population, opening up new possibilities.

Holland's algorithm can be summarized as follows:

Generate a population of chromosomes at random.
Repeat the following steps:
    Evaluate every chromosome's performance.
    Discard the worst performing chromosomes.
    Pair the surviving chromosomes at random.
    Mate every pair to produce two children chromosomes.
    Add the children to the population.
    Randomly alter a small number of chromosomes.
Stop repeating after a fixed number of generations.
Output the best performing chromosome.

In one of his books, the renowned biologist Richard Dawkins reported using a genetic algorithm to reveal a secret message. Dawkins used chromosomes directly as guesses for the secret message. He performed selection by comparing each chromosome to the secret message. Dawkins' *fitness function* is unclear but most likely his program calculated a score for each chromosome. Perhaps the correct letter in the correct place was given a score of +2 points, the correct letter in the wrong place was worth +1, and an incorrect letter was worth 0. The chromosomes with the lowest scores were discarded. The highest scoring chromosomes were 'mated' by means of *crossover*. Crossover generates chromosomes for offspring by exchanging fragments of the parents' chromosomes. A random position is selected within the chromosome. The first child's chromosome is a copy of the father's chromosome up to that point and a copy of the mother's thereafter. The second child's chromosome is a copy of the mother's up to the same point and a duplicate of the father's beyond it. For example, if the parents' chromosomes are:

ABCDEF, LMNOPQ,

and the crossover point is the third letter, the children's letter sequences are:

ABCOPQ, LMNDEF.

Dawkins ran his genetic algorithm in a computer. Initially, the chromosomes were random letter sequences. After ten generations, he reported that the highest scoring chromosome in the population was:

MDLDMNLS ITPSWHRZREZ MECS P.

After twenty, it was:

MELDINLS IT ISWPRKE Z WECSEL.

After thirty:

METHINGS IT ISWLIKE B WECSEL.

After forty-one generations, the algorithm found the secret message:

METHINKS IT IS LIKE A WEASAL.

Unusually, Dawkins' example evaluates the chromosome directly, comparing the letters to the secret message. To solve real-world problems, the chromosome normally controls the construction of a solution, which is then evaluated. For example, Holland used chromosomes to control the behaviour of computer programs simulating commodity markets. Even though they were not initially programmed to do so, his agents evolved so as to contrive speculative bubbles and financial crashes.

NASA employed genetic algorithms to design a radio antenna for one of its space missions. This time, the chromosomes controlled the shape of the antenna. The fitness of the chromosomes was evaluated by calculating the sensitivity of the antenna to incoming radio signals. An evolved X-band antenna flew on the NASA Space Technology 5 mission in 2006.

In 1967, Holland was appointed Professor of Computer Science and Engineering at the University of Michigan. As well as inventing genetic algorithms, he made major contributions to complexity and chaos theory. Very unusually, he also became a Professor of Psychology.

Holland passed away in 2015 at the age of 86. David Krakauer, the President of the Santa Fe Institute, said of him: [112]

> John is rather unique in that he took ideas from evolutionary biology in order to transform search and optimization in computer science, and then he took what he discovered in computer science and allowed us to rethink evolutionary dynamics. This kind of rigorous translation between two communities of thought is a characteristic of very deep minds.

While genetic algorithms remain popular, they are often not the most efficient way to solve a given optimization problem. They work best when there is little understanding of how good solutions are put together. Driven by random permutations, genetic algorithms blindly search the design space. Since they are easy to program, researchers often just let the computer do the work with a genetic algorithm, rather than spend precious time inventing a new fast search algorithm.

While Holland was working on his genetic algorithms in Michigan, the seeds of a revolution in computing were being sown by an obscure agency within the US Department of Defense. Spurred by the Cold War and guided by two visionaries, America began to interlink its computer networks. The endeavour was to have far reaching implications. Along the way, it sparked a new alchemy—the fusion of algorithms with electronics.

7

The Internet

It seems reasonable to envision [...] a 'thinking centre' that will incorporate the functions of present-day libraries.

The picture readily enlarges itself into a network of such centres, connected to one another by wide-band communication lines and to individual users by leased-wire services. In such a system, the speed of the computers would be balanced, and the cost of the gigantic memories and the sophisticated programs would be divided by the number of users.

JCR LICKLIDER
Man-Computer Symbiosis, 1960 [113]

On 4 October 1957, the Soviet Union launched the world's first artificial satellite into Earth orbit. Sputnik 1 was a radio transmitter wrapped up in a metal sphere just 60 cm in diameter. Four radio antennas were fixed to the globe's midriff. For twenty-one days, Sputnik broadcast a distinctive and persistent 'beep-beep-beep-beep' signal. The signal was picked up by radio receivers worldwide as the satellite passed overhead. Sputnik was a sensation. Suddenly, space was the new frontier and the Soviet Union had stolen a march on the West.

US President Dwight D. Eisenhower resolved that the US should never again take second place in the race for technological supremacy. To this end, Eisenhower established two new governmental agencies. He charged the National Aeronautics and Space Administration (NASA) with the exploration and peaceful exploitation of space. Its malevolent sibling—the Advanced Research Projects Agency (ARPA)—was instructed to fund the development of breakthrough military technologies. ARPA was to be the intermediary between the military and research-performing organizations.

The Cold War beckoned.

ARPANET

Nineteen sixty-two saw the founding of ARPA's Information Processing Techniques Office (IPTO). The IPTO's remit was to fund research and development in information technology (i.e. in computers and software). The Office's inaugural director was JCR (Joseph Carl Robnett) Licklider (Figure 7.1).

Licklider (born 1915) hailed from St. Louis. Universally liked, everyone called him 'Lick' for short. Lick never lost his Missouri accent. At college, he studied an unusual cocktail of physics, mathematics, and psychology. He graduated with a PhD in Psychoacoustics, the science of the human perception of sound, from the University of Rochester. He then worked at Harvard University before joining MIT as an Associate Professor. It was at MIT that Licklider first became interested in computers. It transpired that he had a talent, even a genius, for technical problem solving. He carried his new-found passion into his role at ARPA.

Licklider wrote a series of far-sighted papers proposing new computer technologies. *Man-Computer Symbiosis* (1960) suggested that computers should work more interactively with users, responding to their requests in real-time rather than by batch print-out. He proposed the creation of an *Intergalactic Computer Network* (1963) enabling integrated operation of multiple computers over great distances. In *Libraries of the Future* (1965), he argued that paper books should be replaced by electronic devices that receive, display, and process information. He jointly published a paper in 1968 that envisaged using networked computers as person-to-person communication devices. In a ten-year burst of creativity, Licklider predicted personal computers, the Internet, eBooks, and email. His imagination far outpaced reality. His writings set out grand visions for others to chase.

The first step was Project MAC (Multiple Access Computing) at MIT. Up to that point computers had a single user. Project MAC constructed a system whereby a single mainframe computer could be shared by up to thirty users working simultaneously. Each user had their own dedicated *terminal* consisting of a keyboard and screen. The computer switched its attention between the users, giving each the illusion that they had a single, but less powerful, machine at their disposal.

Two years after Licklider's departure from ARPA, Bob Taylor was appointed Director of the IPTO (1965). From Dallas, Taylor (born 1932)

Figure 7.1 JCR Licklider, computer network visionary. (*Courtesy MIT Museum.*)

had previously worked for NASA. Like Licklider, Taylor's background was in psychology and mathematics. Unusually, for a senior executive in a research funding body, Taylor didn't hold a doctoral degree. Whereas Licklider was an ideas guy, Taylor had an almost uncanny knack of delivering breakthrough technologies. Taylor was to spend the next thirty years turning Licklider's visions into reality.

Taylor's enthusiasm for computer networking was borne out of simple frustration. Three computer terminals sat in his office at the Pentagon. Each terminal was connected to a different remote computer—one at MIT, another in Santa Monica, and the third at Berkeley. The computers were not interconnected in any way. To pass a message from one machine to another, Taylor had to type it in again on the other terminal. He remembers: [119]

> I said, oh man, it's obvious what to do. If you have these three terminals, there ought to one terminal that goes anywhere you want to go.

At Taylor's behest, IPTO programme manager Larry Roberts compiled a Request For Quotation for construction of a computer network. The new network was to be called the ARPANET. Initially, the ARPANET would link four sites, with a possible expansion to thirty-five. The winning bid was submitted by Bolt, Beranek, and Newman Technologies (BBN) of Cambridge, Massachusetts.

Three months after the Apollo 11 moon landing, on 29 October 1969, Charley Kline sent the first message on the ARPANET. Kline was a student programmer in Leonard Kleinrock's group at the University of California, Los Angeles (UCLA). His intent was to send a 'LOGIN' command to a computer at SRI, 400 miles away. However, the system crashed after the second character was received. As a result, the first message sent on the ARPANET was the inauspicious fragment 'LO'. About an hour later, after a system re-start, Kline tried again. This time, the login worked.

ARPANET was one of the first networks to employ *packet-switching*. The technique was invented independently by Paul Baran and Donald Davies. Baran, a Polish–American electrical engineer, published the idea in 1964 while working for the RAND Corporation. Davies, a veteran of Turing's ACE project, developed similar ideas while working at the NPL in London. It was Davies that coined the terms 'packet' and 'packet-switching' to describe his algorithm. Davies was later part of the team that built the world's first packet-switched network—the small-scale Mark I NPL Network in 1966.

Packet-switching (Figure 7.2) solves the problem of efficient transport of messages across a network of computers. Imagine a network of nine computers that are physically interconnected by means of cables carrying electronic signals. To reduce infrastructure costs, each computer is connected to a small number of others. Directly connected computers are called neighbours, regardless of how far apart they really are. Sending a message to a neighbour is straightforward. The message is encoded as a series of electronic pulses transmitted via the cable to the receiver. In contrast, sending a message to a computer on the other side of the network is complicated. The message must be relayed by the computers in between. Thus, the computers on the network must cooperate to provide a network-wide communication service.

Before packet-switching, communication networks relied on dedicated end-to-end connections. This approach was common in wired telephony networks. Let's say that computer 1 wishes to communicate with computer 9 (Figure 7.2). In the conventional *circuit-switching* scheme, the network sets up a dedicated electrical connection from computer 1 to computer 3, from 3 to 7, and from 7 to 9. For the duration of the message exchange, all other computers are blocked from sending messages on these links. Since computers typically send short sporadic messages, establishing dedicated end-to-end connections in this way makes poor use of network resources. In contrast, packet-switching provides efficient use of network links by obviating the need for end-to-end path reservation.

In packet-switching, a single message is broken up into segments. Each segment is placed in a packet. The packets are independently

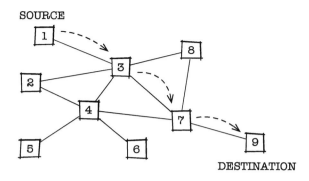

SOURCE

DESTINATION

SOURCE	DESTIN- ATION	PACKET NUMBER	PAYLOAD
1	9	1	This message
1	9	2	is split over
1	9	3	three packets.

Figure 7.2 Packet-switching on a small network.

routed across the network. When all of the packets have been received at the destination, a copy of the original message is assembled. The network transfers packets from source to destination in a series of *hops*. At every hop, the packet is transmitted over a single link between two computers. This means that a packet only blocks one link at a time. Thus, packets from different messages can be interleaved—one after another—on a single link. There is no need to reserve an entire end-to-end connection.

The downside is that links can become *congested*. This happens when incoming packets destined for a busy outgoing link must be delayed and queued.

A packet-switched data network is similar to a road network. The packets comprising a single message are akin to a group of cars making their way to a destination. The cars don't need the entire path be reserved in advance. They simply slot in when the road is free. They may even take different paths to the destination. Each car wends its own way through the network as best it can.

A packet contains a header, a portion of the message (the *payload*), and a trailer (Figure 7.2). The header consists of a start of packet marker, the unique ID of the destination, a message ID, and a sequence number. The sequence number is the packet count within the message (1, 2, 3, …). The sequence numbers are used by the destination computer to assemble the payloads in the correct order. The trailer contains error checking information and an end-of-packet marker.

Computers on the network continuously monitor their connections for incoming packets. When a packet is received, it is processed by the computer. First of all, the ID of the packet's destination is read from the header. If the receiving computer is not the final destination, the device re-transmits the packet on the link offering the fastest route to that destination. The link to be used is determined by means of a *routing table*. The routing table lists the IDs of all computers, or groups of computers, on the network. For each destination, it records the outgoing link that provides the fastest path to that machine. The receiving computer forwards the packet to this link. If the receiving computer is the intended destination, the message ID and sequence number are inspected. When all of the packets in a single message have arrived, the computer concatenates their payloads by sequence number so as to recover the original message (see the Appendix for the full algorithm).

Packet-switching is an example of a *distributed* algorithm. Distributed algorithms run on multiple, independent, but co-operating, computers. Every computer performs its own individual task, which contributes in some way to the grand scheme.

An essential component of a packet-switching system is the algorithm used to populate the routing table. The *routing* algorithm determines how best to populate the routing table. It is the routing algorithm that decides which packets go where.

The original ARPANET routing algorithm based its decisions on exchange of path delay information. In addition to a routing table, every computer maintains a *delay table* (Figure 7.3). The delay table lists every computer ID on the network and, for each machine, the estimated time taken to deliver a packet to it from that point. Periodically, all computers send their delay table to all of their neighbours. On receipt of a neighbour's delay table, a computer adds the time it would take to get a packet to the neighbour. The updated table thus contains a list of every computer ID and the time it would take to get a packet there via the

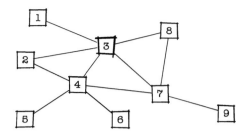

Figure 7.3 The routing table for computer 3 listing the packet destination, the delay via computers 1, 2, 4, 7, and 8, and the ID for the neighbour giving the shortest route from computer 3 to the destination. Delay is measured in hops.

neighbour. The computer compares these times with those already in the routing table. The current entry is updated if the new path is faster. In summary:

For every neighbouring computer:
  Get its delay table.
  For every destination in the table:
    Add the delay from this computer to the neighbour.
    If the new delay is shorter than the delay in the routing table, then:
      Store the link ID and the new delay in the routing table.

The delay of a path can be measured as the number of hops to the destination. Alternatively, the number of packets queued along the path can be counted. This latter option provides an estimate of the congestion on the path and typically yields better results.

One of the benefits of packet-switched networks is robustness to failure. If a link fails, or becomes overly congested, computers can detect the resulting increase in packet queue length and adjust their routing tables so as to avoid the bottleneck. The disadvantage of packet-switching is that packet delivery times are unpredictable. Packet-switching is a 'best effort' service.

After two years at ARPA, the networking visionary, Licklider joined IBM. Later he returned to MIT. Following a second term as IPTO Director, Licklider again resumed his teaching and research career at MIT. He finally retired in 1985.

Bob Taylor went on to found Xerox's Palo Alto Research Center's (PARC's) Computer Science Laboratory in 1970. During his tenure, PARC built a succession of remarkable real-world prototypes. The Center developed a low-cost, packet-switched networking technology for interconnecting computers within a single building. That technology—Ethernet—is now the most common wired computer networking standard in the world. The Lab also invented the point-and-click *graphical user interface*, which was to become ubiquitous in Apple's Mac computers and in the Microsoft Windows operating system. In one of the greatest corporate miscalculations of the twentieth century, Xerox failed to capitalize on PARC's breakthroughs. Taylor left PARC in 1983, later establishing a research centre for Digital Equipment Corporation. He retired in 1996 and passed away in 2017.

The original four-node packet-switched ARPANET grew slowly but steadily with, on average, one *node* (computer site) being added per month. The ARPANET was not the world's first, nor only, computer network. Nevertheless, it was to be the progenitor of the largest computer network on the planet: the Internet.

Internetworking

The first public demonstration of the ARPANET took place at the International Computer Communication Conference (ICCC) in Washington in October 1972. Organized by Bob (Robert) Kahn of BBN, the demo connected forty plus computers. Hundreds of previously sceptical industry insiders saw the demo and were impressed. Packet-switching suddenly seemed like a good idea after all.

Kahn was from New York (born 1938). He graduated with MA and PhD degrees from Princeton before joining BBN. Shortly after the ICCC demo, he moved to the IPTO to oversee further development of the network. Kahn reckoned that the next big challenge was not adding more computers to the ARPANET, but connecting the ARPANET to other networks.

The ARPANET was all a bit samey—the nodes and links used similar fixed line technologies. Kahn imagined a hyperconnected world in which messages would travel transparently between devices on all kinds of networks—wired, radio, satellite, international, mobile, fixed, fast, slow, simple, and complex. This was a glorious concept, to be sure. The question was how to make it work? Kahn came up with a concept he called *open-architecture networking*. The approach seemed promising but the devil was in the details. In 1973, Kahn visited the Stanford lab of ARPANET researcher Vint Cerf and announced: [126]

I have a problem.

Vint (Vinton) Cerf was born in New Haven, Connecticut in 1943. He graduated Stanford University with a BS degree in Mathematics before joining IBM. A few years later, he chose to enrol in graduate school at UCLA. It was at UCLA that Cerf began to work on the ARPANET. Here too he met Kahn for the first time. Shortly after Kahn's ICCC demo, Cerf returned to Stanford as a professor.

Working together, Kahn and Cerf (Figure 7.4) published an outline solution to the inter-networking problem in 1974. In it, they proposed an overarching *protocol* that would be common to all networked computers. The protocol defined a set of messages and associated behaviours that would allow computers to communicate across technologically

Figure 7.4 Inventors of TCP/IP: Vint Cerf (left, 2008) and Robert Kahn (right, 2013). (*Left: Courtesy Vint Cerf. Right:* By Вени Марковски | *Veni Markovski - Own work | CC BY-SA 3.0,* https://commons.wikimedia.org/w/index.php?curid= 26207416.)

disparate networks. A protocol is an agreed sequence of messages that allows computers to interact. For example, in the human world, we use 'Hello' to initiate a conversation and 'Goodbye' to finish it. Everyone knows what the standard messages mean and what should happen next.

As they incorporated feedback from other networking aficionados, Cerf and Kahn's outline snowballed into a detailed technical specification. That specification was to become the first description of the Transmission Control Protocol / Internet Protocol (TCP/IP). Cerf and Kahn believed that TCP/IP would solve the problem of interconnecting distinct computer networks.

The first trial of TCP/IP came on 27 August 1976. A computer installed in an SRI delivery truck, affectionately nicknamed 'the bread van', parked south of San Francisco communicated with a fixed line ARPANET node in the University of Southern California. The demonstration proved that TCP/IP allowed message exchange between radio and wired networks.

The following year, communication across three networks was tested. This time, nodes in Norway and the United Kingdom were internetworked with sites in the US via a transatlantic satellite link. To Cerf, this three-network test was the real thing—proper internetworking using TCP/IP. The experiment on 22 November 1977 heralded the dawn of the Internet.

Today, TCP/IP is the protocol by which computers transmit and receive messages on the Internet. To the user, the most visible element of TCP/IP is the global naming convention for computer nodes. A computer's Internet Protocol *address* uniquely identifies it on the network. An IP address consists of four numbers separated by dots. For example, the IP address of Google's search page is 172.217.11.174. Later, textual names for nodes were added for convenience. The *domain name system* converts these textual names to numeric IP addresses. Thus, the domain name google.com translates to the IP address 172.217.11.174.

In 1983, the ARPANET's original communications protocol—NCP—was replaced by TCP/IP. Around the same time, the ARPANET's routing algorithms were upgraded. Dijkstra's route-finding algorithm began to be used for packet-routing (see Chapter 6). Ironically, the algorithm for finding the shortest path between cities is now far more frequently used to determine the fastest route for data packets traversing the Internet.

ARPANET was formally decommissioned in 1990. By then, it was just another network within a global collection of interconnected networks. The backbone of the Internet in the US was now NSFNET. ARPANET's demise went largely unnoticed. Licklider, the original computer networking visionary, passed away that same year.

Ex-BBN-employee Bob Kahn went on to found the Corporation for National Research Initiatives (CNRI) in 1986. CNRI conducts, funds, and supports research on forward-looking information technologies. He and Vint Cerf founded the Internet Society in 1992 to promote TCP/IP. Cerf continued his work on the Internet, accumulating a lengthy list of corporate and not-for-profit appointments. Perhaps the most eye-catching is his Chief Internet Evangelist role at Google. NASA also engaged Cerf as a consultant on its plans for the Interplanetary Internet. Cerf and Kahn were the recipients of the ACM Turing Award for 2004.

TCP/IP enabled the interworking of a global patchwork of distinct computer networks. There is no central controller on the Internet. So long as computers adhere to the TCP/IP protocols and naming conventions, they can be added to the network. In the period 2005 to 2018, the Internet had 3.9 billion users—more than half of the world's population. This figure looks set to increase further.

Thanks to TCP/IP, humanity has never been so interconnected.

Fixing Errors

Communication systems such as the Internet are designed to transfer exact copies of messages from the transmitter to the receiver. To achieve this, the data in the packet is converted to an electronic signal that is sent from the transmitter to the receiver. The destination device converts the received signal back into data. Often, the received signal is contaminated by electronic *noise*. Noise is the general term for any unwanted signal that corrupts the intended signal. Noise can arise from natural sources or interference from nearby electronic equipment. If the noise is strong enough relative to the signal, it can lead to errors in conversion of the signal back into data. Obviously, errors are not desirable. An error in a sentence might be tolerated. However, an error in your bank balance would not be, unless it happened to be in your favour! For this reason, communications systems incorporate error detection and correction algorithms.

The simplest way to check for, and correct, an error is repetition. To ensure correct delivery, an important piece of data might be sent three times in succession. The receiver compares all three copies. If they match, then it can be assumed that no error occurred. If just two copies match, then the odd-one-out can be assumed to be in-error. The two matching copies are taken to be correct. If no copies match, then the true message is unknown and retransmission must be requested.

For example, if the received error protected message is

HELNO, HFLLO, HELLP

then it is probable that the original was HELLO since the three Hs match, two Es out-vote one F, the next three Ls match, two Ls beat one N, and two Os out-vote one P.

Repetition works well. However, it is very inefficient. If every packet is sent three times, then the number of packets that can be transmitted per second is a third of what it was before error protection was added.

Checksums are much more efficient. The idea is that all characters (letters, digits, and punctuation) are converted to numbers. These numbers are added together and the total—the checksum—is transmitted together with the message. On receiving the packet, the computer recalculates the checksum and compares the result of this computation with the checksum included in the packet. If the calculated and received checksums match, then, most likely, the received packet is error-free. Of course, it is possible that two equal, and opposite, errors occurred in the data or that the checksum and data suffer exactly the same errors. However, these eventualities are extremely unlikely. Most of the time, a checksum mismatch indicates that a transmission error has occurred.

For example, converting HELLO to integers gives:

8 5 12 12 15.

Adding these values produces a checksum of 52. The packet sent is then:

8 5 12 12 15 - 52.

If transmission is error-free, the receiver obtains an exact copy of the message. It recalculates the checksum as 52. This matches the checksum in the packet, and all is well.

What if the first character is mistakenly received as an F?

$\boxed{6}$ 5 12 12 15 - 52.

This time, the calculated checksum (50) does not match the checksum at the end of the packet (52). The receiver knows that an error has occurred.

Checksums are common. For example, the International Standard Book Number (ISBN) at the front of this book contains a checksum. All books printed after 1970 have an ISBN that uniquely identities the title. Current ISBNs are thirteen digits long and the last digit is a check digit. The check digit allows computers to verify that the ISBN has been typed in or scanned correctly.

A basic checksum merely detects errors. It is impossible to work out which number is incorrect. The error could even be in the checksum itself. Ironically, in this case, the message itself is correct. Basic checksums require retransmission of the message to fix an error.

Richard Hamming (Figure 7.5) wondered if checksums could be modified to provide error correction, as well as detection.

Born in Chicago in 1915, Hamming studied Mathematics, ultimately attaining the PhD degree from the University of Illinois. As the Second World War ended, Hamming joined the Manhattan Project in Los Alamos. He worked, as he put it, as a 'computer janitor', running calculations on IBM programmable calculators for the nuclear physicists.[130] Disillusioned, he moved to Bell Telephone Laboratories in New Jersey. Bell Labs was the research wing of the burgeoning Bell Telephone Company founded by Alexander Graham Bell, the inventor of the telephone. In the late 1940s and 1950s, Bell Labs employed a stellar cast of communications researchers. Hamming was in his element:[130]

Figure 7.5 Richard Hamming, inventor of error-correcting codes, 1938. (*Courtesy of the University of Illinois Archives, 0008175.tif.*)

We were first-class troublemakers. We did unconventional things in un-conventional ways and still got valuable results. Thus, management had to tolerate us and let us alone a lot of the time.

Renowned for his bad jokes, not everyone appreciated Hamming's way of doing things: [130]

He is very hard to work with because he does a lot of broadcasting and not a lot of listening.

Hamming's interest in error correction stemmed from his own frustrations in dealing with an unreliable relay-based computer. All too often, he would leave a program running over the weekend only to discover on the Monday morning that the job had failed due to a fault in the computer. Exasperated, Hamming wondered why the machine couldn't be programmed to find and fix its own errors.

The simplest form of checksum is the *parity bit*. Modern electronic computers process information as binary numbers. Unlike decimal numbers, *binary* numbers only use two digits: zero and one. Whereas in decimal every column is worth ten times the one to its right, in binary every column is worth double. Thus, in binary—moving right-to-left from the fractional point—we have the units, twos, fours, eights, sixteens, and so on. For example:

$$1011 \text{ binary} = (1 \times 8) + (0 \times 4) + (1 \times 2) + (1 \times 1) = 11 \text{ decimal.}$$

Similarly, counting from zero to fifteen in binary gives the sequence:

0, 1, 10, 11, 100, 101, 110, 111,
1000, 1001, 1010, 1011, 1100, 1101, 1110, 1111.

Thus, four binary digits, or *bits*, can represent the decimal integers from zero to fifteen.

A parity bit can be appended to a binary number for the purposes of error detection. The value of the parity bit (0 or 1) is selected such that the total number of 1 bits, including the parity bits, is even. For example, the data word:

0 1 0 0 0 1

is protected by appending a parity bit with the value zero:

0 1 0 0 0 1 - 0.

This keeps the number of 1 bits even (i.e. two 1-bits).

As with checksums, the parity bit is sent along with the *data word*, that is, the sequence of data bits. To check for errors, the receiver simply counts the number of 1s. If the final count is even, then it can be assumed that no error occurred. If the count is odd, then, in all likelihood, one of the bits suffered an error. A 0 has been mistakenly flipped to a 1, or vice versa. For example, an error in bit two gives the word:

$$0 \boxed{0} 0\,0\,0\,1 - 0.$$

This time, there is an odd number of 1 bits, indicating that an error has occurred.

In this way, a single parity bit allows detection of a single error. If two bits are in error, then the packet appears valid, but is not. For example:

$$\boxed{1}\,\boxed{0}\,0\,0\,0\,1 - 0$$

seems to be correct since there is an even number of 1s. As a consequence, additional parity bits are needed when there is a high error rate.

Hamming devised a clever way to use multiple parity bits to detect and correct single bit errors. In Hamming's scheme, every parity bit protects half of the bits in the word. The trick is that no two parity bits protect the same data bits. In this way, every data bit is protected by a unique combination of parity bits. Hence, if an error occurs, its location can be determined by looking at which parity bits are affected. There can only be one data bit protected by all of the parity bits showing errors.

Let's say that a data word containing eleven data bits is to be transmitted:

$$1\,0\,1\,0\,1\,0\,1\,0\,1\,0\,1.$$

In Hamming's scheme, eleven data bits require four parity bits. The parity bits, whose value is to be determined, are inserted into the data word at positions which are powers of two (1, 2, 4, and 8). Thus, the protected word becomes:

$$?\,?\,1\,?\,0\,1\,0\,?\,1\,0\,1\,0\,1\,0\,1,$$

where the question marks indicate the future positions of the parity bits. The first parity bit is calculated over the bits at odd numbered positions (numbers 1, 3, 5, etc.). As before, the value of the parity bit is selected to ensure that there is an even number of 1s within the group. Thus, the first parity bit is set to 1:

①:①:⓪1⓪:①0①0①0①.

The circles indicate the bits in the parity group. The second parity bit is calculated over the bits whose positions, when written in binary, have a 1 in the twos column (2, 3, 6, 7, etc.):

1⓪①:0①⓪:1⓪①01⓪①.

The third parity bit is calculated over bits whose positions, in binary, have a 1 in the fours column (4, 5, 6, 7, 12, etc.):

1 0 1①⓪①⓪:1 0 1⓪①⓪①.

The fourth parity bit is calculated over bits whose positions, in binary, have a 1 in the eights column (8, 9, 10, 11, etc.):

1 0 1 1 0 1 0⓪①⓪①⓪①⓪①.

This then is the final protected data word, ready for transmission.

Now, imagine that the protected data word suffers an error at bit position three:

1 0 |0| 1 0 1 0 0 1 0 1 0 1 0 1.

The receiver checks the word by counting the number of ones in the four parity groups:

①0⓪1⓪1⓪0①0①0①0① = 5 ones;
1⓪⓪10①⓪01⓪①01⓪① = 3 ones;
1 0 0①⓪①⓪0 1 0 1⓪①⓪① = 4 ones;
1 0 0 1 0 1 0⓪①⓪①⓪①⓪① = 4 ones.

The first and second parity groups both show errors (i.e. they have an odd numbers of ones). In contrast, the third and fourth groups do not indicate errors (i.e. they have an even numbers of ones). The only data bit that is in the first and second groups and is not in the third and fourth groups is bit three. Therefore, the error must be in bit number three. The error is easily corrected by flipping its value from 0 to 1.

Hamming's ingenious algorithm allows detection and correction of single errors at the cost of a small increase in the total number of bits sent. In the example, four parity bits protect eleven data bits—just a thirty-six per cent increase in the number of bits. Hamming codes are remarkably simple to generate and check. This makes them ideal for high-speed processing, as required in computer networks, memory, and storage systems. Modern communication networks employ a mixture of Hamming codes, basic checksums, and newer, more complex, error-correcting codes to ensure the integrity of data transfer. A mistake in your bank balance is extremely unlikely.

After fifteen years in Bell Labs, Hamming returned to teaching, taking up a position at the Naval Postgraduate School in Monterey, California. Hamming received the Turing Award in 1968 for his codes and other work on numerical analysis. He died in 1998 in Monterey, just one month after finally retiring.

One of the great flaws of the Internet is that it was not designed with security in mind. Security has had to be grafted on afterwards, with mixed results. One of the difficulties is that packets can be easily read *en route* by eavesdroppers using electronic devices. *Encryption* circumvents eavesdropping by altering a message in such a way that only the intended recipient can recover the original text. An eavesdropper might still intercept the altered text, but the scrambled message will be meaningless.

Until the end of the twentieth century, encryption algorithms were intended for scenarios in which the encryption method could be agreed in absolute secrecy before communication took place. One imagines a queen furtively passing a top-secret codebook to a spy at a clandestine rendezvous. However, this approach doesn't translate well to computer networks. How can two computers secretly exchange a codebook when all data must be sent over a vulnerable public network? At first, it seemed that encryption wasn't at all practical in the brave new world of the computer network.

Secret Messages

Encryption was used in ancient Mesopotamia, Egypt, Greece, and India. In most instances, the motivation was secure transmission of military or political secrets. Julius Caesar employed encryption for important personal letters. The Caesar Cipher replaces every letter in the original text with a substitute letter. The substitute letter is a fixed number of places away from the original in the alphabet. To make patterns more difficult to spot, Caesar's Cipher strips away spaces and changes all

letters to capitals. For example, a right shift by one place in the alphabet
leads to the following encryption:

Hail Caesar

IBJMDBFTBR.

The As become Bs, the Es change to Fs, and so on. Any Zs would be
replaced by As since the shift wraps around the end of the alphabet.
The encrypted message—the *ciphertext*—is sent to the receiver. The
receiver recovers the original message—the *plaintext*—by shifting every
letter one place left in the alphabet. The Bs become As and so on
returning the original 'HAILCAESAR' message. Thanks to the patterns
in natural language, the missing spaces are surprisingly easy to infer.

Traditional encryption methods, such as the Caesar Cipher, rely on
an algorithm and a secret *key*. The key is a piece of information that is
essential for successful encryption and decryption. In the Caesar Cipher,
the key is the shift. The algorithm and the key must be known to the
sender and the intended recipient. Typically, security is maintained by
keeping the key secret.

No encryption scheme is perfect. Given sufficient time and a clever
attack, most codes can be broken. The Caesar Cipher can be attacked by
means of frequency analysis. An attacker counts the number of times
that each letter of the alphabet occurs in the ciphertext. The most
commonly occurring is probably the substitute for the vowel E, since
E is the most common letter in the English language. Once a single
shift is known, the entire message can be decrypted. Almost all codes
have vulnerabilities. The question is, 'How long does the attack take to
perform?' If the attack takes an unacceptably long period of time, then
the cipher is secure from a practical point of view.

In computer networks, key distribution is problematic. The only
convenient way to pass a key is via the network. However, the network is
not secure against eavesdropping. Sending a secret key via the Internet
is equivalent to making it public. How could a sender and a receiver
possibility agree on a secure key if all they can do is send public messages?
The question became known as the Key Distribution Problem. The first
glimmer of a solution came from a group working at Stanford in the
early 1970s.

Martin Hellman was born in New York in 1945. He studied Electrical
Engineering at New York University before moving to California to

study for his MSc and PhD degrees at Stanford. After stints at IBM and MIT, Hellman returned to Stanford in 1971 as an Assistant Professor. Against the advice of his peers, Hellman started to work on the Key Distribution Problem. Most thought it foolhardy to expect to find something radically new—something that the well-resourced US National Security Agency (NSA) had missed. Hellman was unperturbed. He wanted to do things differently to everyone else. In 1974, Hellman was joined in the hunt for a solution by Whitfield Diffie.

From Washington, DC, Diffie (born 1944) held a degree in Mathematics from MIT. After graduation, Diffie worked programming jobs at MITRE Corporation and his alma mater. However, he was captivated by cryptography. He struck out to conduct his own independent researches on key distribution. On a visit to IBM's Thomas J. Watson Laboratory in upstate New York, he heard about a guy called Hellman who was working on similar stuff at Stanford. Diffie drove 5,000 miles across the US to meet the man who shared his passion. A half-hour afternoon meet-up extended long into the night. A bond was formed.

The duo was joined by PhD student Ralph Merkle. Born in 1952, Merkle had previously come up with an innovative approach to the Key Distribution Problem while studying as an undergraduate at the University of California, Berkeley.

In 1976, Diffie and Hellman published a paper describing one of the first practical algorithms for public key exchange. The paper was to revolutionize cryptography. The myth that all keys had to be private was shattered. A new form of coding was born: *public key cryptography*.

The Diffie–Hellman–Merkle key exchange scheme showed that two parties could establish a secret key by means of public messages. There was a hitch, though. Their method required the exchange and processing of multiple messages. As a result, the algorithm was not ideal for use on networks. However, their paper did suggest an alternative.

Traditional encryption algorithms use a *symmetric* key, meaning that the same key is used for encryption and decryption. The drawback with symmetric encryption is that the key must be keep secret at all times. This requirement creates the Key Distribution Problem.

In contrast, public key encryption uses two keys: an encryption key and a different—*asymmetric*—decryption key. The pair of keys must meet two requirements. First, they must work successfully as an encryption–decryption pair, i.e. encryption with one and decryption with the other must return a copy of the original message. Second, it must be impossible to determine the decryption key from the encryption key. Therein lies the beauty of public key cryptography. If the decryption key cannot be determined from the encryption key, then the encryption key can be made public. Only the decryption key needs to be keep secret. Anyone can use the public encryption key to send a secret message to the private key holder. Only the recipient who possesses the private decryption key can decipher and read the message.

Imagine that Alice wants to be able to receive encrypted messages (Figure 7.6). She creates an asymmetric key pair by means of a key generation algorithm. She keeps the decryption key to herself. She publicly releases the encryption key on the Internet. Let's say that Bob wants to send a secret message to Alice. He obtains Alice's encryption key from her Internet posting. He encrypts the message using Alice's encryption key and sends the resulting ciphertext to Alice. On receipt, Alice decrypts the ciphertext using her private decryption key. In short:

Alice generates the encryption and decryption key pair.
Alice keeps the decryption key to herself.
She makes the encryption key public.
Bob encrypts his message using Alice's public encryption key.
Bob sends the encrypted message to Alice.
Alice decrypts the encrypted message using her secret
decryption key.

The scheme works beautifully with one proviso. There must be no way to determine the private decryption key from the public encryption key. Therein lay the difficulty. No one knew how to create asymmetric keys where the decryption key could not be worked out from the encryption key. What was needed was a *one-way function*—a computation whose input could not be easily inferred from its output. If such a function could be found, its output could be the basis of the public key and its input the basis of the private key. There would be no way to reverse the computation. Attackers would be unable to recover the decryption key.

Diffie and Hellman's paper described public key encryption but did not offer a one-way function. The concept was inspired, but they couldn't find a way to make it work.

At MIT's Laboratory for Computer Science in Boston, Ronald Rivest read Diffie and Hellman's paper with mounting excitement. There and then he undertook to hunt for a suitable one-way function. A function that would unlock public key encryption. Rivest persuaded two friends

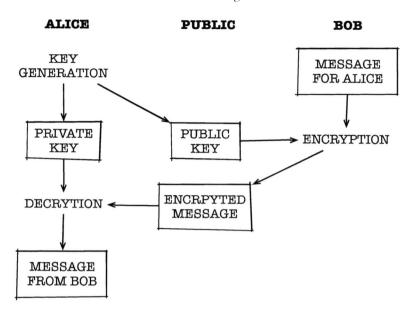

Figure 7.6 Public key cryptography.

and colleagues, Adi Shamir and Leonard Adleman, to help him in the search. All three held Bachelor's degrees in Mathematics and PhDs in Computer Science. Rivest (born 1947) hailed from New York state, Adi Shamir (1952) was a native of Tel Aviv, Israel, and Adleman (1945) grew up in San Francisco. The impromptu team spent a year generating promising ideas for one-way functions only to discard each and every one. None were truly one-way. Perhaps there was no such thing as a one-way function.

The trio spent Passover 1977 as guests at a friend's house. When Rivest returned home, he was unable to sleep. Restless, his mind turned to the one-way encryption problem. After a while, he hit upon a new function that might just work. He wrote the whole thing down before dawn. The next day, he asked Adleman to find a flaw in his scheme—just as Adleman had done for every other suggestion. Curiously, Adleman couldn't find a weakness. The method seemed to be robust to attack. They had found a one-way function. Rivest, Shamir, and Adleman's algorithm for key generation was published later that same year. It quickly became known as the RSA algorithm from its inventors' initials. RSA is now the cornerstone of encryption on the Internet.

RSA encryption and decryption are reasonably straightforward. The encryption key consists of two numbers—a modulus and an encryption exponent. The decryption key also contains two numbers—the same modulus and a unique decryption exponent.

To begin, the original textual message is converted to a sequence of numbers. Encryption is applied to groups of numbers as follows:

Calculate the input number to the power of the encryption exponent.
Calculate the remainder after dividing this number by the modulus.
Output the remainder.

To decipher:

Calculate the number received to the power of the decryption exponent.
Calculate the remainder after dividing by the modulus.
Output the remainder.

Let's say that the encryption key is (33, 7), the message is 4, and the decryption key is (33, 3). Calculating 4 to the power of 7 (that is, 4 multiplied by itself 7 times) gives 16,384. The remainder after dividing 16,384 by 33 is 16. So, 16 is the ciphertext.

To decipher, calculate 16 to the power of 3 giving 4,096. The remainder after dividing 4,096 by 33 is 4. The output, 4, is the original message.

How does the scheme work? The process hinges on *clock arithmetic*. You may have come across the *number line*—an imaginary line with the integers marked off evenly along it, much like a ruler. Starting at zero, the number line stretches off into infinity. Now imagine that there are only 33 numbers on the line (0–32). Roll up this shortened line into a circle. The circle looks much like the face of an old-fashioned wall clock marked with the numbers 0 to 32.

Imagine starting at zero and counting around the clock. Eventually, you come to 32 and after that the count goes back to 0, 1, 2 and so on. You keep going around and around the clock face.

Clock arithmetic mirrors the effect of the remainder operation. Dividing 34 by 33 gives a remainder of 1. This is the same as going around the clock once and taking one more step.

In the example, encryption moves the clock hand 16,384 steps around the clock face. In the end, the clock hand is left pointing to 16: the ciphertext. Decryption starts at 0 and moves the clock hand 4,096 steps around the clock face. In the end, the clock points to 4: the original message.

The encryption and decryption keys are complementary. The key pair is especially selected so that one exponent undoes the effect of the other. The number of complete rotations around the clock face does not matter. In the end, all that matters is the digit that the hand is pointing to.

The key pair is produced by means of the RSA key generation algorithm. This is the heart of the RSA encryption. The first two steps contain the one-way function:

Pick two large prime numbers with similar values.
Multiply them to get the modulus.
Subtract one from each prime number.
Multiply the results to give the *totient*.
Choose the encryption exponent as a prime number between 1
    and the totient.
Repeat the following steps:
    Choose a constant value.
    Calculate the constant multiplied by the totient and add
        one.
    Calculate this number divided by the encryption exponent.
Stop repeating when the result is a whole number.
Let the decryption exponent equal the whole number.
Output the encryption key (modulus and encryption
    exponent).
Output the decryption key (modulus and decryption
    exponent).

The algorithm is complicated, but let's try an example. Say we start with the prime numbers 3 and 11. These are too small to be secure from attack, but they will do for now. The modulus is then $3 \times 11 = 33$. The totient is $(3-1) \times (11-1) = 20$. We select 7 as the encryption exponent since it is a prime number between 1 and the totient. Taking a constant value of 1, we have $1 + 1 \times 20 = 21$. The decryption exponent is then equal to 21 divided by 7, that is, 3. Thus, the encryption key is $(33, 7)$ and the decryption key is $(33, 3)$.

Attacking the key pair boils down to finding the two prime numbers that were multiplied together to give the modulus. Multiplication disguises the selected primes. For large numbers, there are many pairs of prime numbers that might have been multiplied together to give the modulus. An attacker would have to test a huge number of primes to crack the code. When a large modulus is used, brute-force search for the original prime numbers is prohibitively time-consuming.

The other steps in the key generation algorithm ensure that the encryption and decryption are reciprocal. That is, decryption undoes encryption for all values between zero and the modulus.

A *Scientific American* article introduced the RSA algorithm to a general readership in 1977. The item concluded with a $100 challenge to crack an RSA ciphertext given the encryption key. The modulus was reasonably large—129 decimal digits. It took seventeen years to crack the code. The winning team comprised six hundred volunteers leveraging spare time on computers all over the world. The plaintext turned out to be the distinctly uninspiring: [136]

The Magic Words are Squeamish Ossifrage

An ossifrage is a bearded vulture. The $100 prize worked out at 16 cents per person. The point was proven. RSA is military-grade encryption.

Public key encryption is now built into the Secure Socket Layer of the World Wide Web. When a web site address is preceded by "https:", your computer is using SLL and, with it, the RSA algorithm to communicate with the remote server. At last count, seventy per cent of web site traffic used SSL. Over the years, the length of keys has had to increase to prevent ciphertexts being cracked by the latest supercomputers. Today, keys containing 2,048 or more bits (617 decimal digits) are commonplace.

The academics had proven the naysayers wrong. Governmental electronic espionage agencies could be beaten at their own game. At least, that's how it seemed.

Amidst the hullabaloo over RSA, the Head of the US National Security Agency stated publicly that the intelligence community had known about public key encryption all along. This raised eyebrows. Was his claim factual, bravado, or just plain bluff? Diffie's curiosity was piqued. He made discrete enquiries and was directed to look across the Atlantic towards GCHQ. Government Communication Headquarters is the UK's electronic intelligence and security agency. During the Second World War, it was the body that oversaw the code-breaking work at Bletchley Park. Diffie wrangled two names from his contacts: Clifford Cocks and James Ellis. In 1982, Diffie arranged to meet Ellis in a pub in Cheltenham. A GCHQ stalwart to the last, the only hint that Ellis gave was the cryptic: [137]

You did a lot more with it than we did.

In 1997, the truth came out. GCHQ published the papers of Cocks, Ellis, and another actor, Malcolm Williamson. Among the documents was a history of events at GCHQ written by Ellis. In it, he refers to Diffie–Hellman–Merkle's 'rediscovery' of public key cryptography.

In the early 1970s, James Ellis hit upon the idea of public key cryptography. His name for the technique was 'non-secret encryption'. An engineer by profession, Ellis couldn't come up with a satisfactory one-way function. Clifford Cocks happened to hear about Ellis's discovery over a cup of tea with Nick Patterson. Cocks, an Oxford and Cambridge educated mathematician, found himself at a loose end that evening. He decided to study the one-way encryption problem. Spectacularly, Cocks solved the problem that evening. Just like the RSA team, he settled on multiplication of two large prime numbers. This was four years ahead of Rivest, Shamir, and Adleman. Cocks circulated the idea in the form of an internal GCHQ paper. Malcolm Williamson picked up on the memo and added a missing piece on key exchange some months later.

The codebreakers at GCHQ couldn't find a flaw in Ellis, Cocks, and Williamson's unconventional method. Nevertheless, the hierarchy remained unconvinced. Non-secret encryption languished in the office drawers at GCHQ. Gagged by the Official Secrets Act, the authors said nothing. They watched on the sidelines as the Stanford and MIT teams reaped the glory. Ellis never gained the satisfaction of public recognition. He passed away just a few weeks before the embargo on his papers was lifted.

Rivest, Shamir, and Adleman won the ACM Turing Award for 2002. Diffie and Hellman were honoured as the recipients of the 2015 award.

Meanwhile, the Internet grew and grew. By 1985, there were 2,000 hosts (computer sites) on the Internet. Most were owned by academic institutions. While data transport worked well, the networking programs of the 1980s were unappealing. Their user interface was text-heavy, monochrome, and cumbersome to use. To reach a wider audience, computing needed a makeover.

8

Googling the Web

A memex is a device in which an individual stores all books, records, and communications, and which is mechanised so that it may be consulted with exceeding speed and flexibility.

[The essential feature of the memex] is a provision whereby any item may be caused at will to select immediately and automatically another.

VANNEVAR BUSH
The Atlantic, 1945[140]

By the 1970s, minicomputers were well established in scientific institutes, universities, and large companies. Similar in height and girth to an American-style refrigerator, a minicomputer was far cheaper than the antiquated mainframe, but was still a hefty purchase. While effective for large-scale data processing, the machines were not user-friendly. User terminals consisted of a monochrome monitor and a chunky keyboard. A fixed grid of green letters and numbers illuminated a black background. An arcane textual command was required for every action.

In 1976, two California kids—Steve Jobs and Steve Wozniak—launched the first pre-assembled microcomputer onto an indifferent market. For the first time, a commercial computer was small enough to fit on a desktop. Even better, the Apple I was cheap enough that it could be bought by, and used by, just one person. Apple followed up with an improved offering the next year. Despite the compact dimensions of the hardware, sales were muted until the arrival of VisiCalc.

VisiCalc was the world's first commercial spreadsheet program. VisiCalc allowed users to enter text, numbers, and formulae into on-screen tables. The secret sauce was that VisiCalc would automatically perform the calculations specified in the formulae when numbers were entered into the spreadsheet. There was no need to write a program to get a few calculations done. Suddenly, business users could play with their sales figures without resorting to the IT Department. People bought the Apple II just to use VisiCalc. The desktop computer market burgeoned.

IBM spotted the disruption late. Digital Equipment Corporation (DEC) and IBM were the leading suppliers of minicomputers. In a catch-up play, IBM launched their own Personal Computer in 1981. Supported by the corporation's vast sales network, the IBM PC was a commercial success.

Three years later, Apple beat IBM to the punch once more. Apple released the first graphical user interface (GUI) for an affordable computer. Apple's attractively designed Macintosh computer shipped with a keyboard, high-resolution screen, and—radically—a mouse. The mouse allowed users to control the computer by clicking on icons and menus. Programs now ran side by side in adjustable windows. The mouse was a hit. Old-fashioned textual commands were for geeks.

Although Apple commercialized the GUI, the technology was invented elsewhere. The mouse came from Douglas Engelbart at SRI. The GUI was created at Xerox PARC under Bob Taylor's watch. Steve Jobs happened to see a demo of PARC's GUI and had to have one for his own products.

The Macintosh left Apple's competitors scrambling. IBM's software partner hastily added a GUI into their next operating system, Microsoft Windows. At the same time, IBM's PC sales came under pressure from low-cost copycat, or *clone*, hardware manufacturers. IBM and Apple simply couldn't compete on price. Soon, cheap PC clones running Windows were on every desktop in every enterprise.

While the GUI made working with a single computer much easier, accessing data via networks remained a chore. The Internet provided global connectivity for specialist computer centres. However, networking programs still used textual commands. Even worse, every remote access system (bulletin boards, library catalogues, remote logins, etc.) had its own peculiar set of commands. Using the Internet was incredibly frustrating. Easy-to-use software for sharing data between computers was sorely needed. Surprisingly, the fix didn't come from the computer industry. It came from a European particle physics laboratory.

World Wide Web

Tim Berners-Lee was born in London in 1955. He graduated Oxford University with a degree in Physics. Following in the footsteps of his parents—who had been programmers on the Ferranti Mark I—Berners-Lee became a professional software developer. Between industry jobs, he worked as a contractor at CERN for six months in 1980. Four years later, he returned to work on the design of CERN's computer network.

CERN is the European Organisation for Nuclear Research. The body is based at a sprawling campus in Geneva, Switzerland. In 1984, Berners-Lee was one of 10,000 staff, students, and visiting researchers working on a myriad of loosely connected projects. The place was a hodgepodge of organizational structures, vested interests, cultures, and languages. Coordination and communication were next to impossible. It seemed to Berners-Lee that networked computers could assist in the day-to-day task of information sharing.

Berners-Lee proposed a scheme whereby a user's desktop computer could download and view electronic *pages* stored on remote, or *server*, computers. Each page would be held as a data file which could be transferred over the Internet. The data file would include text and special formatting tags specifying how the page was to be displayed. A piece of software called a *browser* running on the user's computer would request the remote page from the server and display it. Each page would be identified by a unique name consisting of the server's ID followed by the file name. Later, a prefix indicating which protocol to use was added. The complete identifier is now called the page's uniform resource locator (URL).

A key feature of Berners-Lee's proposal was that the pages would contain *hyperlinks*. A hyperlink, or *link* or short, is a piece of text, or an image, that is tagged with a reference to another web page. When a user selects the link, their browser automatically displays the referenced page. Links greatly simplified navigation between pages. Click on a link and the associated page appears.

Hyperlinks seemed startlingly modern. In fact, the idea wasn't new at all. Hyperlinks were first suggested by Vannevar Bush in a visions-of-the-future article published back in 1945. What was new was that networked computers and software could turn Bush's speculation into reality.

Berners-Lee and Robert Cailliau put together a detailed document describing the system. Emphasizing the global reach of the Internet and the hyperlinked connectivity of pages, they named the proposed system the WorldWideWeb (WWW). Spaces were later added to make the name more readable. The document described two elements: a formal definition of the file format (i.e. the allowed contents) of web pages; and a *protocol* specifying the messages and behaviours that the browser and server software should use to communicate.

Within a year of receiving approval for the project, Berners-Lee finished the software for the first WWW browser and server. The world's first web site went live on 6 August 1991. The original pages are still available at:

info.cern.ch/hypertext/WWW/TheProject.html

At Berners-Lee's insistence, CERN released the WWW specifications and software for free. Uptake was sluggish. In 1993, a team from the University of Illinois at Urbana-Champaign led by Marc Andreessen released a new web browser. Mosaic was compatible with Berners-Lee's server software but, crucially, ran on Microsoft Windows. Windows-based PCs were far more common than the obscure workstations that Berners-Lee had programmed at CERN. By the end of the year, 500 web servers were online. Sites on particle physics and computing were joined by pages devoted to financial news, web comics, movie databases, maps, webcams, magazines, company brochures, restaurant advertisements, and pornography.

The following year, as the WWW gained traction, Tim Berners-Lee left CERN to found and lead the World Wide Web Consortium (W3C). The W3C operates as a not-for-profit body working in collaboration with industry partners to develop and promote the WWW. The Consortium is the guardian of the World Wide Web standards to this day. True to Berners-Lee's ideals, the WWW standards have remained open, free, and publicly available. Anyone can build a compatible web browser or server without having to seek permission or pay a royalty.

The WWW provided the world with a low cost, robust, and easy-to-use platform for sharing information. It was up to website developers to figure out what to do with it.

Amazon Recommends

In the year that Berners-Lee left CERN, a Wall Street investment banker happened upon a startling statistic. Spurred by the popularity of the Mosaic browser, web usage had grown 2,300 per cent year-on-year. The number was ludicrous. Double digit growth is hard to come by, nevermind four digit. The banker found lots of information on the web but almost nothing for sale. Surely, this was an untapped market. The question was: 'What to sell?'

At the time, the Internet was way too slow to stream music or videos. Product delivery would have to be by the US postal service. An online store would be like a mail-order business—only better. Customers could view an up-to-date product catalogue and place orders via the web. The banker looked up a list of the top twenty mail-order businesses. He concluded that book retail would be a perfect fit. The investment banker had stumbled upon the opportunity of a lifetime.

At thirty years old, Bezos was DE Shaw & Company's youngest ever Senior Vice President (VP). Born in Albuquerque, New Mexico, Bezos grew up in Texas and Florida. He attended Princeton University, graduating in Computer Science and Electrical Engineering. After college, he worked a series of computer and finance jobs, quickly climbing the ladder to Senior VP.

Bezos's brainwave left him with a dilemma. Should he pack in his six-figure New York banking job to go selling books? Bezos called on an algorithm that he employed for making life altering decisions like this:[143]

> I wanted to project myself forward to age 80 and say, 'OK, I'm looking back on my life. I want to minimize the number of regrets I [will] have.'
>
> I knew that when I was 80, I was not going to regret having tried this.
>
> I knew the one thing I might regret is not ever having tried. [That] would haunt me every day.

Bezos quit his Wall Street job and embarked on a mission to build a bookstore in a place that didn't really exist—online.

He needed two things to get his Internet business off the ground: staff with computer skills and books to sell. Seattle on the northwest coast of the US had both. The city was home to Microsoft and the country's largest book distributors. He and his wife of one year—MacKenzie Bezos (née Tuttle)—boarded a plane to Texas. On arrival, they borrowed a car from Bezos' dad and drove the rest of the way to Seattle. MacKenzie took the steering wheel while Bezos typed up his business plan on a laptop computer.

Using his parents' life savings as seed capital, Bezos set up shop in a small two-bedroom Seattle house. On 16 July 1995, the amazon.com website went live.

Figure 8.1 Greg Linden, designer of the first Amazon recommender system, 2000. (*Courtesy Greg Linden.*)

Sales were healthy. Amazon re-located to 2nd Avenue. The offices were so over-crowded that Greg Linden (Figure 8.1), one of the new hires, was forced to work in the kitchen.

Linden was on a time-out from the University of Washington. He loved the buzz of the start-up but fully intended to return to college to do a PhD in Computer Science. In the meantime, Linden reckoned that he could help Amazon sell more books. He was convinced that a *recommender system* would help turn Amazon home page views into book purchases.

A product recommender system analyses customers' purchasing decisions and suggests products that they might like to buy. A reader that has previously bought a collection of crime novels might well be tempted into buying a Sherlock Holmes special edition or a Raymond Chandler novel. Presenting these books as suggestions to the user might well encourage them to buy one.

In essence, recommendation is advertising. The distinction is that recommendation is *personalized*. It is tailored to the individual according their interests. Linden's hope was that personalized recommendations would increase the number of sales per home page view in comparison to traditional one-size-fits-all advertisements. A handful of experimental recommender engines were already online. In contrast, Amazon's recommender would have to work at scale in a commercial setting. Linden pitched his idea to management and was given the go-ahead to build Amazon's first recommender system.

Linden based his algorithm on a simple intuition. If a pair of products are commonly bought together then it is likely that a customer who already owns one will buy the other. The items don't have to have been originally purchased as a pair. All that matters is that users often buy both. The reason why doesn't matter, either. The books could be by same author. They might be from the same genre. Perhaps they are both study guides for the same state examinations. In terms of recommending products, the timing of, and rationale, for the pairing are irrelevant. All that matters is that customers often buy both items.

Linden's algorithm records all of the purchases made on the Amazon website. When a user checks out, the algorithm notes the purchaser's unique ID and the names of the books purchased. The algorithm recovers a list of all previous purchases made by that user. It then pairs the new items with all of the user's previous purchases.

Imagine that Mary buys *Charlotte's Web* and has previously purchased *The Little Prince* and *Pinocchio*. This creates two new book pairings:

Charlotte's Web & *The Little Prince*
Charlotte's Web & *Pinocchio*

This information is used to update a *product similarity table*. The table lists all of the books on the Amazon web site down the rows and across the columns. Thus, every pair of books has two entries in the body of the table. The entries record the number of times that a pair of books (row and column) have been purchased by a single user. This number is the *similarity score* for the pair.

Mary's purchase of *Charlotte's Web* generates two new pairs. As a result, the entries for *Charlotte's Web* and *The Little Prince* are incremented by one, as are the entries for *Charlotte's Web* and *Pinocchio* (Table 8.1).

When a user arrives at the Amazon web site, the algorithm uses the similarity table to generate recommendations. The method begins by retrieving the user's purchasing history. For every book that the user has bought, the algorithm looks up the corresponding row in the similarity table. The algorithm scans across these rows looking for non-zero entries. For each, the algorithm looks up the book title at the top of column. These book names and similarity scores are recorded in a list. On completion, the list is reviewed, and any duplicates or items already bought by the user are removed. The remaining items are sorted by similarity score and the books with the highest scores are presented to the user as recommendations.

In summary, the algorithm works as follows:

Take the similarity table and user's purchasing history as input.
Create an empty list.
Repeat for every item in the purchasing history:
    Find the matching row in the similarity table.
    Repeat for every column in that row:
        If the similarity score is greater than zero,
        then add the matching title and score to the list.
    Stop repeating at the end of the row.
Stop repeating at the end of the purchasing history.
Remove any duplicates from the list.
Remove any book that the user has already bought.
Sort the list by similarity score.
Output the titles with the highest similarity score.

Let's say that *Charlotte's Web* and *The Little Prince* have been pair purchased by four different users; *Charlotte's Web* and *Pinocchio* were bought by one customer; as were *The Little Prince* and *Pinocchio* (Table 8.1). Imagine that Nicola web surfs to the Amazon site. The recommender algorithm recovers her purchasing history and finds that she has only bought one book so far: *The Little Prince*. The algorithm scans along the *The Little Prince* row in the similarity table to find two non-zero entries: 4 and 1. Looking up the corresponding column headers, the algorithm discovers that *Charlotte's Web* and *Pinocchio* are paired with *The Little Prince*. Since *Charlotte's Web* has the higher similarity score (4), it is presented to Nicola as the best recommendation. In other words, Nicola is more likely to buy *Charlotte's Web* than *Pinocchio* since, in the past, more customers have bought both *The Little Prince* and *Charlotte's Web*.

Table 8.1 Similarity table for three books. The entries indicate the number of times that each pair of books has been purchased by a user.

	Charlotte's Web	The Little Prince	Pinocchio
Charlotte's Web	—	4	1
The Little Prince	4	—	1
Pinocchio	1	1	—

Recommendation accuracy improves significantly with data set size. The more data in the product similarity table and the more extensive a user's history, the better the recommendations get. With more data, oddities disappear, and major trends emerge. Given enough data, recommender algorithms are surprisingly accurate. According to a recent McKinsey report, thirty-five per cent of Amazon's sales come from product recommendation.

The year that Amazon launched (1995), the WWW boasted forty-four million users and twenty-three thousand sites. The following year the number of users doubled and the number of websites rose by a factor of ten. The World Wide Web was becoming a frenzy.

The sheer number of webpages began to create problems for users. Users mindlessly *surfed* the web, clicking on one link after another. How could anyone find what they were looking for? Manually curated website *directories* that sought to catalogue the WWW by topic were swamped. Finding material online became increasingly slow and frustrating. Users desperately wanted a website that would magically display exactly the right link.

Google Web Search

In the spring of 1995, Sergey Brin was asked to show a new guy around Stanford University. Aged twenty-one, Brin had been a student at Stanford for two years. Born in Moscow, Brin's parents migrated to the US when he was six. He graduated with a degree in Computer Science and Mathematics from the University of Maryland at just nineteen before winning a scholarship to attend graduate school at Stanford.

The new guy was twenty-two-year-old Larry Page. Page had recently finished a degree in Computer Science at the University of Michigan. Hailing from the Midwest, Page was overawed by Stanford. He half expected to be sent home.

Brin and Page took a near instant dislike to one another. Page:[149]

I thought he was pretty obnoxious. He had really strong opinions about things, and I guess I did, too.

Brin:

We both found each other obnoxious.

Nonetheless, the two grad students found common ground in a love of repartee:

We had a kind of bantering thing going.

Page wasn't sent home on the next bus. He started his PhD. He hung out with Brin. They duo began to collaborate on projects of mutual interest.

Everyone in computer science knew that the WWW was hot. Netscape had just floated on the stock market at a valuation of $3 billion despite having zero profits. The company's only product was a web browser, which they gave away for free. Netscape's Initial Public Offering (IPO) heralded the beginning of the Wall Street *dotcom* bubble. The popular dotcom moniker derived from the URL extension assigned to commercial web sites. Web start-up after web start-up hit crazy stock market valuations.

Page's thoughts turned to the problem of web search. Website directories clearly weren't the way to go. Drilling down through alphabetic lists of categories, then topics, then sub-topics was way too time consuming for users. Queries made more sense. Just type in what you are looking for, and the most relevant links should appear. The difficulty was that matching query terms with the titles of web sites didn't work at all well. Most of the links returned were either irrelevant or low quality, or both. Users were forced to sift through a lot of rubbish before they found what they were looking for. Page realized that accurately ranking weblinks based on their importance and relevance to the query was key. An accurate website ranking system would push useful links to the top of the list. The question was how to accurately rank websites?

Page was familiar with an existing paper-based system for ranking academic research papers. Research papers are typically ranked according to the number of times that they have been referenced in other publications. The idea is that the more often a paper is referenced, or cited, the more important it is. Page realized that hyperlinks are a lot like references. A hyperlink from one page to another indicates that the author of the linking page thinks that the linked page is, in some way, important or relevant. Counting the number of links directed to a page might be a good way to assess the page's importance. Working from this insight, Page developed an algorithm for ranking the importance of web pages. In a play on his own name, he dubbed the method PageRank.

PageRank does more than simply count citations: it takes into consideration the importance of the pages that link to the page being rated. This stops webmasters artificially raising the rank of a page by creating spurious pages linking to it. To have any impact, the linking pages must, themselves, be important. In effect, PageRank ranks webpages based on the communal intelligence of website developers.

Every webpage is allocated a PageRank score: the higher the score, the more important the page. The score for a page is equal to the sum of the weighted PageRanks of the pages that link to it plus a *damping term*.

The PageRank of an incoming link is weighted in three ways. First, it is multiplied by the number of links from the linking page to the page being scored. Second, it is *normalized*, meaning that it is divided by the number of links on the linking page. The rationale is that a hyperlink from a page that contains many links is worth less than a hyperlink from a page with a small number of links. Third, the PageRanks are multiplied by a *damping factor*. This damping factor is a constant value (typically 0.85) which models the fact that a user might jump to a random page rather than follow a link. The damping term compensates for this by adding one minus the damping factor to the total (usually 0.15).

PageRank can be thought of as the probability that a web surfer who selects links at random will arrive a particular page. Lots of links to a page mean that the random surfer is more likely to arrive there. Many hyperlinks to those linking pages also means that the surfer is more likely to arrive at the destination page. Thus, its PageRank depends not only on the number of links to a page but also on the PageRank of those linking pages. Therefore, the PageRank of a destination web-page depends on all of the links that funnel into it. This funnelling effect goes back one, two, three, and more links in the chain. This dependency makes PageRank tricky to calculate. If every PageRank depends on every other PageRank, how do we start the calculation?

The algorithm for computing PageRanks is iterative. Initially, the PageRanks are set equal to the number of direct incoming links to a page divided by the average number of inward links. The PageRanks are then recalculated as the total of the weighted incoming PageRanks plus the damping term. This gives a new set of Page Ranks. These values are then used to calculate the PageRanks again, and so on. With every iteration, the calculation brings the PageRanks closer to a stable set of values. Iteration stops when the PageRanks show no further change.

Imagine a mini-WWW comprised of just five pages (Figure 8.2). At first glance, it is difficult to determine which page is most important. Two pages have three incoming links (B and E). One page has two inward links (A). The other two pages are less popular, with just one incoming link (C and D).

Calculation of the PageRank scores begins by creating a table showing the number of links between pages (Table 8.2). Every row corresponds to a page, as does every column. The entries in the table record the number of links from the page in the row to the page in the column. Thus, reading across a row, we have the number of outgoing links for the named page. Scanning down a column, we have the number of incoming links for the named page.

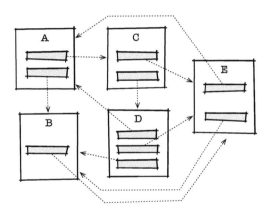

Figure 8.2 Five inter-linked web pages.

Table 8.2 Table showing the number of links between pages. The table also includes the total number of links. A page may not link to itself.

	To A	To B	To C	To D	To E	Total out
From A	–	1	1	0	0	2
From B	0	–	0	0	1	1
From C	0	0	–	1	1	1
From D	1	1	0	–	1	3
From E	1	1	0	0	–	2
Total in	2	3	1	1	3	

The algorithm creates a second table listing the PageRanks for every page (Table 8.3). To start with, the algorithm populates the table with a rough estimate of the PageRanks. This is the number of incoming links for a page divided by the average number of inward links.

The algorithm then recalculates the PageRanks page by page, that is, column by column in the PageRank table. A page is processed by calculating the incoming weighted PageRanks for a single column in the links table. The PageRank of each incoming link is obtained by looking up the current PageRank for that page (Table 8.3). This value is multiplied by the number of incoming links from the page. The result is divided by the total number of outgoing links on the page (Table 8.2). This value is multiplied by the damping factor (0.85) to obtain the weighted PageRank. This calculation is performed for all incoming links. The resulting weighted incoming PageRanks are totalled and added to the damping term (0.15). This gives the new PageRank for the page. This value is appended to the PageRank table.

The PageRank calculation is repeated for all of the pages. When all of the pages have been processed, the new PageRanks are compared to the previous values. If the change is small, the process has converged and the results are output. Otherwise, the calculation is repeated (the complete algorithm is listed in detail in the Appendix).

For example, imagine performing the second iteration in calculating the PageRank of page A. Its PageRank is the sum of the weighted PageRanks of the pages with incoming links, that is, pages D and E. The initial PageRank of D is 0.5. The number of links from D to A is 1. The number of links from D is 3. Therefore, the weighted PageRank from D to A is $0.5 \times 1 \div 3 = 0.167$. Similarly, the weighted PageRank from E to A is $1.5 \times 1 \div 2 = 0.75$. Including the damping term, the new PageRank for A is $0.167 + 0.75 + 0.15 = 1.067$.

The iterative procedure balances the PageRanks so that they reflect the inter-connection of the webpages. The larger values flow to the

Table 8.3 Table showing the final PageRank values obtained for the mini-WWW example.

Iteration	A	B	C	D	E
1	1.00	1.50	0.50	0.50	1.50
5	0.97	1.38	0.56	0.40	1.69

more interconnected pages, the smaller to the less well connected. Eventually, the numbers settle down to a steady-state that balances the flow.

For the mini-WWW example, the PageRank scores converge after about five iterations (Table 8.3). In the end, webpage E has the greatest PageRank. Page E has the same number of incoming links as B but ranks ahead of it. This is because page B, which links to E, only has a single link. Whereas, page E, which links to B, has two links. Normalization reduces the PageRank of page B.

To try out his algorithm, Page, in collaboration with Brin and Stanford Professor Rajeev Motwani, built a prototype search engine. They used a *web crawler* to download a summary of the WWW. A web crawler explores the WWW much like a human surfer, following link after link. Along the way, the crawler keeps a snapshot of every web page that it comes across. Starting with a few manually curated URLs, a crawler quickly creates a local summary of the WWW. The PageRank algorithm is applied to the resulting dataset to score the importance of the pages. On receiving a user query, the team's search engine—BackRub—searches the local summary for matching page titles. The resulting list of pages is sorted by PageRank and displayed to the user.

BackRub was only ever used internally within Stanford. Nonetheless, its performance was sufficiently promising that Page and Brin decided to expand the service.

They bought more computers so that a larger portion of the WWW could be indexed, and a greater number of concurrent queries handled. On reflection, they decided BackRub was a poor name. They needed something that intimated the future scale of the WWW. They decided on 'Googol'—a googol being a one followed by one hundred zeros. By mistake, they wrote the word down as 'Google'. Brin drew up a multicoloured logo for the new site. By July 1998, the Google search engine had indexed twenty-four million pages. Crucially, Google's search results were light years ahead of their competitors'. The new kid on the block was ready for the big time.

In August, Brin and Page were introduced to Andy Bechtolsheim. A Stanford alumnus, Bechtolsheim had cofounded two successful tech start-ups. Brin and Page pitched the Google proposition to Bechtol-sheim. He liked what he heard. Right there and then he wrote a cheque for $100,000 made out to Google, Inc. No negotiations, no terms and conditions, and no valuations. Bechtolsheim just wanted to be part of it. In spirit, he felt he was paying back an old favour granted by one of his early backers. Clutching the cheque, Brin and Page neglected to point out that Google, Inc., didn't exist—yet.

The following February, *PC Magazine* reported that the fledgling Google search engine: [153]

> has an uncanny knack for returning extremely relevant results.

One year later, Google, Inc., received $25 million in funding from venture capitalists Sequoia Capital and Kleiner Perkins. Sequoia Capital and Kleiner Perkins were, and still are, Silicon Valley royalty. Their names on the investor sheet were almost worth more than the money.

The following year, Google launched AdWords. AdWords allows advertisers to bid for their links to be listed on the Google search page alongside the PageRank results. The promoted links are clearly delineated from the PageRank results so that users can distinguish between PageRank results and advertisements. AdWords proved to be much more effective than traditional advertising. This was hardly surprising, as AdWords was offering products that customers were already actively searching for. Companies flocked to the new advertising platform. AdWords turned free web search into a goldmine.

The PageRank algorithm was patented by Stanford University. Google exclusively licensed the algorithm back from Stanford for use in its search engine in exchange for 1.8 million shares. In 2005, Stanford University sold its Google shares for $336 million. That transaction probably makes PageRank the most valuable algorithm in history.

The Dotcom Bubble

Hot on the heels of the Netscape IPO in 1995, investments poured into web start-ups. Profits didn't seem to matter. The only metric for valuation was the number of users that a site could claim. Web investment became a pyramid scheme. The tech-loaded NASDAQ index of shares quintupled in value between 1995 and 2000. The index peaked at 5,048 on 10 March 2000. Thereafter, it plummeted. On 4 October 2002, the NASDAQ index returned to 1,139, shedding more than three quarters of its value. It was almost back to where is started in 1995. The dotcom bust hit the tech industry hard. A slew of web start-ups went to the wall. It would be fifteen long years before the NASDAQ recovered its year 2000 peak.

The speculative dotcom bubble disguised steady and continuous growth in Internet usage. Fuelled by an expanding user base, the surviving web companies grew rapidly.

Linden's product recommender algorithm was widely emulated. He left Amazon in 2002. After spells at two start-ups and Google, he now works at Microsoft. Today, Amazon sells everything 'from A to Z'. In 2019, Amazon overtook Walmart as the world's largest retailer. According to *Forbes* magazine, Jeff Bezos—Amazon founder and CEO—became the world's richest person in 2018. His net worth was estimated at $156 billion.

Google IPOed in 2004 at a valuation of $23 billion. The company was a mere six years old. In 2016, Google's parent company—Alphabet—was valued at almost half a trillion dollars. As of writing (2019), Brin and Page are among the top ten richest people in the US according to *Forbes* magazine. Their estimated personal wealth is $35–40 billion, each.

Tim Berners-Lee was granted a knighthood by the Queen of England in 2004 in recognition of his achievements. He is holder of the 2016 ACM Turing Award. *The Richest* website estimates his net worth at $50 million. A small fortune, but a far cry from the wealth of the web billionaires.

When the dotcom bubble burst (2002), there were roughly half a billion active Internet users. The web afforded the world unprecedented access to information, online shopping, and entertainment. However, a nineteen-year-old student living in a college dorm was convinced that what people really wanted was gossip. That insight, combined with a lot of hard work, would make him a multibillionaire.

9

Facebook and Friends

With all these data you should be able to draw some just inference.

SHERLOCK HOLMES to Dr WATSON
Sir ARTHUR CONAN DOYLE
The Sign of Four, 1890[160]

Mark Zuckerberg (Figure 9.1) was born in White Plains, New York, in 1984. His father taught him how to program when he was in middle school. Later, his dad hired a professional programmer to tutor him. A whiz in high school, Zuckerberg was always destined for the Ivy League. He enrolled in Harvard University and selected joint honours in Computer Science and Psychology.

As well as a love of coding, Zuckerberg had an abiding interest in the behaviour of people. He realized early on that most people are fascinated by what other people are doing. This obsession lies at the very heart of everyday gossip, weighty biographies, celebrity culture, and reality TV. At Harvard, he began to experiment with software that would facilitate the basic human need to connect and interact with others.

Zuckerberg set up a website called Facemash.[162] Facemash was modelled on an existing web site by the name of Hot Or Not.[161] Both sites displayed side-by-side images of two male, or two female, students and asked the user to select the more attractive of the two. Facemash collated the votes and displayed ranked lists of the 'hottest' students. Controversially, Facemash used photographs of students downloaded from Harvard web sites. The site was popular with a certain cohort of students but upset a lot of others. According to the on-campus newsletter, the site landed Zuckerberg in front of a disciplinary board.[163]

Afterwards, Zuckerberg turned to constructing a new website for *social networking*. A few such sites already existed, allowing users to share information about themselves. Most previous sites were aimed at people

Figure 9.1 Facebook co-founder Mark Zuckerberg, 2012. (*By JD Lasica from Pleasanton, CA, US - Mark Zuckerberg, | Wikimedia Commons | CC BY 2.0, https://commons. wiki-media.org/w/index.php?curid=72122211. Changed from colour to monochrome.*)

seeking dates. Instead, Zuckerberg wanted his social network to help Harvard students communicate. The idea was that users would enter personal profiles and post news. This wasn't the sort of news that would make the headlines but it was the sort of chit-chat that students loved. To sidestep university rules, the new site required that users upload their own data. The new site—thefacebook.com—was launched in February 2004. Zuckerberg was nineteen.

Facebook's News Feed

Word about Facebook spread quickly. Four days after launch, it had 450 registered users. Students used the site for all sorts of things—planning parties, organizing study sessions, and, inevitably, dating. Gradually, Zuckerberg opened the social network up to students from other US universities. In June, he was offered $10 million for the site. He wasn't remotely interested. As the number of users grew, Zuckerberg took on investment, began hiring, and dropped out of college.

In the beginning, the only way to find a fresh post was to check users' profile pages for updates. Mostly, checking pages was a waste of time— there was just nothing new to see. Zuckerberg realized that it would be helpful for users to have a page summarizing the latest posts from their pals.

Over the next eight months, the Facebook News Feed algorithm was born. The algorithm proved to be the biggest engineering challenge that the young company had faced. News Feed wasn't just a new feature. It was a re-invention of Facebook.

The idea was that News Feed would produce a unique news page for every single user. The page would list the posts most relevant to that particular user. Everyone's feed would be different – personalized for them by the system.

Facebook activated News Feed on Tuesday, 5 September 2006. User reaction was almost unanimous. Everyone hated it. People felt it was too stalker-esque. In fact, nothing was visible that had not been available on Facebook previously. However, News Feed made it easier to see what was going on in everyone else's lives. It seemed that Zuckerberg had misjudged users' emotional reaction to a perceived change in their data privacy.

Anti-News Feed groups sprang up on Facebook and flourished. Ironically, students were using the very feature that they were objecting to, to help them protest. For Zuckerberg, this was proof positive that News Feed had worked. The numbers backed him up. Users were spending more time on Facebook than ever before. Facebook made its apologies, added privacy controls, and waited for the fuss to die down.

The engineering challenge at the core of News Feed lay in creating an algorithm that would select the best news items to display to a user. The question was: how could a computer algorithm possibly determine what a human user was most interested in? The details of Facebook's News Feed algorithm remain a closely guarded secret. However, some information was disclosed in 2008.

The original News Feed algorithm was called EdgeRank. The name seems to have been a nod to Google's PageRank. Every action on Facebook is called an edge, be it a user post, a status update, a comment, a *like*, a group join, or an item share. An EdgeRank score is calculated for every user and every edge by multiplying three factors:

$$EdgeRank = affinity \times weight \times time\ decay.$$

Affinity is a measure of the user's degree of connection to the edge. It indicates how close the user is to the person who created the edge. Friends on Facebook are considered closer than non-friends. The more mutual friends that two friends have, the greater their affinity. The number of interactions between users affects their affinity score. For example, affinity is boosted if users often comment on each other's posts. Affinity wanes over time if users stop interacting with one another.

Weight depends on edge type. Edges that require more effort to create have a higher weight. A comment, for example, has a greater weight than a like.

All else being equal, EdgeRank decays as an edge ages. This ensures that the algorithm prioritizes more recent posts over older items.

Every fifteen minutes, EdgeRanks are re-calculated for every user and every post. A user's NewsFeed is prepared by sorting posts in descending order of their EdgeRank scores. Over time, the relative EdgeRank scores change. One post may be demoted as it ages. Another's rank might rise as it receives a flurry of likes. This dynamism encourages users to keep checking their feeds in the hope of seeing something new.

News Feed popularized *viral messaging*. Users cannot broadcast messages on Facebook. They simply post a message in the hope that it will propagate to their social network. If an item receives lots of attention, in the form of likes or comments, then its EdgeRank score is boosted. As the score increases, the item appears in a wider circle of users' feeds. Popular messages can cross between user communities, much like the spread of a virus.

Facebook parlayed News Feed views to revenue by interspersing user posts with sponsored advertisements. The company was publicly listed on the stock exchange in 2012 at a valuation in excess of $104 billion. The Facebook website and apps now have around 2.4 billion monthly users. Even allowing for *bots* (i.e. programs mimicking users) this is a large proportion of the world's population. As of 2020, Zuckerberg remains CEO of Facebook. According to *Forbes* magazine, Zuckerberg's net worth is estimated at around $60 billion.

News Feed offers a form of personalization, tailoring news content to the individual. While personalization had been available on the web since before the arrival of Linden's Amazon recommender, it was an algorithm developed for Netflix that took the technology to the next level. That algorithm leveraged machine learning to identify and exploit the latent patterns in large volumes of user data. Machine learning combined with big data would soon transform business and science.

The Netflix Prize

Netflix was founded in 1997 by Reed Hastings and Marc Randolph. Originally natives of the east coast, Hastings and Randolph gravitated to Silicon Valley. Both navigated the merger and acquisition maelstrom of the technology industry, becoming serial entrepreneurs. They first met when Hastings' company acquired a software start-up for which Randolph worked. Living close to one another, the colleagues began to carpool. Over the course of their daily commutes, Hastings and Randolph hatched a plan for a new business venture.

The business proposition was straightforward—online movie rental. Subscribers selected movies that they wanted to watch via the company's web site. As they became available, the movies were posted out on disk (DVD). After viewing, customers returned the DVDs by mail.

The service proved popular. Subscribers liked having access to a large movie library, plus they enjoyed the convenience of DVDs arriving on their doorstep. The key to success was ensuring that customers actually enjoyed the movies they received in the post. Following Amazon's lead, Netflix added a recommendation engine to their web site. Netflix's recommender, Cinematch, worked well. Nevertheless, by 2006, the company was looking for something better. Rather than develop another algorithm in-house, Netflix took the unusual step of launching a public competition. The company announced a $1 million prize for the first recommender system that was ten per cent more accurate than Cinematch.

To facilitate the competition, Netflix released a training dataset containing 100 million movie ratings allotted by nearly half a million customers to almost 18,000 different movies. Every data point consisted of the movie's name, the user's name, the star rating assigned (1 to 5), and the date that the rating was logged. The movies' names and ratings were real, but the users' names were anonymized so that individuals could not be identified.

In addition, a second, qualifying, dataset was released. Its contents were similar to the training dataset except that Netflix held back the star ratings. The qualifying dataset was much smaller, containing just 2.8 million entries.

The goal of the competition was to build a recommender that could accurately predict the redacted movie ratings in the qualifying dataset. Netflix would compare the estimates provided by a competitor with the hidden user assigned ratings. The competitor's estimates were evaluated by measuring the *prediction error*—the average difference between the predicted and actual ratings squared.

The $1 million prize attracted hobbyists and serious academic researchers alike. As far as the academics were concerned, the dataset was gold dust. It was very difficult to get hold of real-world datasets of this size. At the outset, most thought that a ten per cent improvement in accuracy was going to be trivial. They underestimated the effectiveness of Cinematch.

Ratings predictions can be made in a large number of ways. The most effective technique used in the competition was to combine as many different predictions as possible. In predictor parlance, any information which can be used as an aid to prediction is a *factor* that must be taken into account in the final reckoning.

The simplest factor is the average rating for the movie in the training dataset. This is the average across all users who have watched that particular movie.

Another factor that can be considered is the generosity of the user whose rating is being predicted. A user's generosity can be calculated as their average rating minus the average across all users for the same movies. The resulting generosity modifier can be added to the average rating for the movie being predicted.

Another factor is the ratings given to the movie by users who generally score in the same way as the user in question. The training dataset is searched for these users. The prediction is then the average of their ratings for the movie.

Yet another factor is the ratings given by the user to similar movies. Again, the training dataset is inspected. This time, movies typically allocated similar ratings to the show in question are identified. The average of the user's ratings for these films is calculated.

These factors, and any others available, are combined by weighting and addition. Each factor is multiplied by a numeric value, or weight. These weights control the relative importance of the factors. A high weight means that the associated factor is more important in determining the final prediction. A low weight means that the factor is of less importance. The weighted factors are summed to give the final prediction.

In summary, the prediction algorithm is then:

Take the training and qualifying datasets as input.
Repeat for every user-movie combination in the qualifying
    dataset:
    Repeat for every factor:
        Predict the user's movie using that factor.
    Stop repeating when all factors have been assessed.
    Weight and sum the factor predictions.
    Output the final prediction for that user and movie.
Stop repeating when all user-movie combinations have been
    predicted.

Imagine that the algorithm is trying to predict Jill's rating for *Toy Story* (Table 9.1). The first factor is simply the average rating for *Toy Story* in the training dataset. This gives an estimate of 3.7 stars. The second factor—Jill's generosity—is obtained by calculating Jill's average rating and subtracting the average rating for the same movies in the training dataset. Jill's average rating is 4 while the average for the same movies is 3.1 Therefore, Jill's generosity bonus is +0.9. Adding this to the global average gives 4.6 stars. Next, the dataset is searched for users whose previous ratings are similar to Jill's. This is clearly Ian and Lucy. They gave *Toy Story* 5 stars, so that's another factor. The fourth factor requires that movies that Jill has watched which normally attain similar ratings to *Toy Story* are found. The obvious ones are *Finding Nemo* and *The Incredibles*. Jill gave these two movies an average rating of 4.5 stars. So, that is another factor. In conclusion, we have four estimates of Jill's rating of *Toy Story*: 3.7, 4.6, 5, and 4.5 stars.

Table 9.1 Movie ratings dataset.

	Toy Story	Finding Nemo	The Incredibles	Frozen
Ian	5	4	4	2
Jill	?	4	5	3
Ken	1	2	1	4
Lucy	5	4	5	2

If the last two factors are typically the most reliable, we might well use the weights: 0.1, 0.1, 0.4, 0.4. Multiplying and summing gives a final prediction of 4.6 stars. Jill should definitely watch *Toy Story*!

Although this general approach was common, teams varied in the specific features they exploited. Sixty features or more were not unusual. Teams also experimented with a wide range of similarity metrics and ways to combine predictions. In most systems, the details of the predictions were controlled by numerical values. These numerical values, or *parameters*, were then tuned to improve the prediction results. For example, the weight parameters were tweaked to adjust the relative importance of factors. Once the factors are identified, accurate rating prediction depends on finding the best parameter values.

To determine the optimum parameter values, teams turned to machine learning (see Chapter 5). To begin with, a team sets aside a subset of the training dataset for *validation*. Next, the team guesses the parameter values. A prediction algorithm is run to obtain predictions for the ratings in the validation dataset. The prediction error is measured for these estimates. The parameters are then adjusted slightly in the hope of reducing the prediction error. These prediction, evaluation, and parameter adjustment steps are repeated many times. The relationship between the parameter values and the prediction error is monitored. Based on the relationship, the parameters are tuned so as to minimize the error. When no further reduction in error can be obtained, training is terminated and the parameter values frozen. These final parameter values are used to predict the missing ratings in the qualifying dataset and the results submitted to Netflix for adjudication.

Overall, the training algorithm works as follows:

Take the training and validation datasets as input.
Guess the best parameter values.
Repeat the following:
>   Run the prediction algorithm for all items in the validation
>   dataset.
>   Compare the predicted ratings to the actual ratings.
>   Adjust the parameters to reduce the error.
Stop repeating when no further improvement is obtained.
Output the parameters which give the minimum prediction
>   error.

The beauty of the machine learning approach is that the computer can experiment with far more parameter combinations than any human.

At first, teams progressed rapidly. The top algorithms quickly hit six, seven, and eight per cent improvements over Cinematch. Then, accuracy stalled. Teams poured over their validation results to find out what was going wrong. The stumbling block was an issue that became known as the *Napoleon Dynamite* problem. *Napoleon Dynamite* was the hardest movie to predict a rating for. Most movie ratings could be predicted fairly easily because there was something similar in the dataset. The trouble with *Napoleon Dynamite* was that there was nothing quite like it. The film was an oddball independent comedy that had become a cult hit. People either loved it or hated it. It was the kind of movie that friends would argue over for hours on end.

Although *Napoleon Dynamite* was the hardest movie to predict, it wasn't alone. There were enough hard-to-predict movie ratings to halt progress. Some competitors began to wonder if a ten per cent uplift was even possible.

Two years after the competition started, a handful of competitors realized that there was a way forward after all. Each team had designed a custom algorithm. Each algorithm had its own particular strengths and weaknesses. In many cases, the strengths and weaknesses of competing algorithms were complementary. Teams realized that accuracy could be enhanced by combining the estimates from multiple algorithms. Again, machine learning was employed to determine the best way to combine the various predictions.

Accuracies began to push upwards once more. The more different the algorithms, the better the combined ensemble results seemed to be. Teams rushed to integrate their algorithms with other high performing solutions. The competition reached its climax in a frenzy of team mergers and assimilations.

On 21 September 2009, the Netflix Prize was won by BellKor's Pragmatic Chaos. The team had achieved a 10.06% improvement on Cinematch. The group was an amalgamation of KorBell out of AT&T Research in the US; Big Chaos from Austria; and Pragmatic Theory from Canada. All in, it was a group of seven people working across the globe, primarily communicating via email. Of course, the merger meant that the $1 million prize had to be split seven ways. Still, it was a pretty good payday. Pity the second placed team who, after three years, lost by just ten minutes.

In a plot twist, Netflix decided not to deploy the winning algorithm. The company had already replaced Cinematch with the winner of an earlier stage. Netflix reckoned that an 8.43% improvement was good enough and left it at that.

The competition was a resounding success. In total, 51,051 contestants from 186 countries organized into 41,305 teams had entered.

While the competition was underway, Netflix announced a shift from disk rental to online streaming of movies across the Internet. The company soon discontinued its disk rental service entirely. Today, Netflix is the world's largest Internet television network, with more than 137 million subscribers. Marc Randolph left Netflix in 2002. He is now on the board of several tech companies. Reed Hastings continues as Netflix CEO. He is on Facebook's Board of Directors as well. Hastings is currently at number 504 on the Forbes Billionaires List (2019).

McKinsey recently reported that a staggering seventy-five per cent of Netflix's views are based on system recommendations. Netflix estimates that personalization and movie recommendation services save the company $1 billion every year through enhanced customer retention.

In 2009, it was starting to seem like big data, coupled with machine learning, could predict almost anything.

Google Flu Trends

That year, an eye-catching paper in *Nature* suggested that data analytics applied to Google web search queries could track the spread of influenza-like diseases across the US. The concept was intuitively appealing. People who feel ill often use Google web search to find out about their symptoms. A spike in user queries related to the flu might well indicate an outbreak of the disease.

It was a significant announcement. Seasonal flu is a major health problem. Every year, there are tens of millions of flu cases worldwide, causing hundreds of thousands of deaths. Furthermore, there is a persistent risk that a new, more virulent, strain of the disease will evolve. The flu pandemic of 1917–1918 killed somewhere between twenty and forty million people—more than died in the First World War.

The authors of the paper employed machine learning to determine the relationship between Google search queries and real-world flu statistics. The flu statistics were provided by the US Centers for

Disease Control and Prevention (CDC). The CDC monitors outpatient visits to hospitals and medical centres across the US. The research team compared regional data on the percentage of flu-related physician visits with Google queries made in the same area over the same time period. The algorithm considered fifty million candidate Google queries comparing the occurrence pattern of each with the observed flu statistics. The researchers identified forty-five search queries whose occurrence patterns closely matched the flu outbreaks in the CDC data. These search queries were then harnessed to predict the flu.

The group evaluated their algorithm by predicting the percentage of flu-related physician visits recorded in a completely different time period. They found that the algorithm consistently, and accurately, predicted the CDC data one to two weeks ahead of time. The study was deemed a success. It appeared that the flu could be predicted by analysing Google searches. To assist medical agencies, Google launched Google Flu Trends, a website to provide real-time estimates of the number of flu cases.

The initiative was widely lauded. Big data and machine learning were delivering real-time medical information at national scale for free. Yet, some researchers were sceptical. There were concerns about aspects of the study. The algorithm had been trained using seasonal flu data. What if the forty-five search queries were associated with the season, rather than the flu itself? The original paper noted that the query 'high school basketball' was well correlated with the CDC flu data. High school basketball season coincides with winter flu season, but high school basketball itself doesn't indicate that anyone really has the flu. The researchers had excluded basketball queries from the system but what if other seasonal correlations were hiding in the data?

A non-seasonal outbreak of influenza virus A(H1N1) provided the first opportunity to test the claims and counter-claims. The epidemic started in the summer, rather than the winter, and came in two waves. This time, the algorithm's predictions did not match the CDC data. In response, the team modified the training dataset to include seasonal and non-seasonal flu events. They also allowed less common search queries. After these modifications, the revised algorithm accurately predicted both waves of the H1N1 epidemic.

All seemed well until a paper appeared in *Nature* in February 2013. The flu season hit early that winter—the earliest since 2003—and was worse than normal, leading to an unusually large number of fatalities, particularly among the elderly. Google Flu Trends overestimated the peak of the CDC data by more than fifty per cent—a huge error. Soon, more Google Flu Trends errors were reported. One research group even demonstrated that Google Flu Trends' predictions were less accurate than basing today's flu predictions on two-week-old CDC data. Amidst the clamour, Google Flu Trends was discontinued.

What went wrong?

With hindsight, the original work used too little CDC data for training. There was so little CDC data and so many queries, that some queries just had to match the data. Many of these matches were random. The statistics happened to match but the queries were not a consequence of anyone actually having the flu. In other words, the queries and flu data were correlated but there was no causal relationship. Second, there was too little variability in the epidemics captured in the training data. The epidemics happened at much the same time of year and spread in similar ways. The algorithm had learned nothing about atypical outbreaks. Any deviation from the norm and it floundered. Third, media hype and public concern surrounding the deaths in 2013 probably led to a disproportionate number of flu queries which, in turn, caused the algorithm to overshoot.

The bottom line is that machine learning algorithms are only as good as the training data supplied to them.

The science of *nowcasting* has moved on significantly since Google Flu Trends. Real-world conditions are now determined at scale and low cost by means of networked data-gathering devices and analytics algorithms. Sentiment analysis applied to Twitter posts is used to predict box office figures and election outcomes. Toll booth takings are used to estimate economic activity in real-time. The motion sensors in smartphones have been monitored to detect earthquakes. Undoubtedly, nowcasting of disease epidemics will be revisited in the future with the aid of more reliable health sensors.

Meanwhile, in 2005, one year before the Netflix Prize was launched, a group of IBM executives were on the hunt for a fresh computing spectacular. The 1997 defeat of world Chess champion Garry Kasparov by IBM's Deep Blue computer had hit headlines around the world. That victory was more about computer chip design than novel algorithms. Nevertheless, the event was a milestone in the annals of the computing. IBM wanted a sequel for the new millennium. The challenge had to be well chosen—a seemingly impossible feat that would grab the general public's attention. Something truly amazing ...

10

America's Favourite Quiz Show

> I, for one, welcome our new computer overlords.
>
> <div align="right">KEN JENNINGS
On <i>Jeopardy</i>, 2011[180]</div>

IBM's T. J. Watson Research Center occupies a futuristic building in Yorktown Heights, New York. The building's low semi-circular facade sweeps away from the entrance to vanishing points at the far ends of the structure. Its three-storey array of black-framed plate glass windows looks out onto wooded parkland. The Center is famous for a succession of historic breakthroughs in both electronics and computing. Incongruously, amidst the snow of early 2011, Watson Center was the venue for a TV game show.

Jeopardy is an institution in the US. The show has aired almost continuously since 1964. In every episode, three contestants compete on the buzzer to win cash prizes. The show's unique feature is that the quizmaster gives the 'answers' and the players provide the 'questions'. In truth, the presenter's 'answers' are cryptic clues. The players' replies are merely phrased as questions. For example, the clue: [182]

> One legend says this was given by the Lady of the Lake and thrown back in the lake on King Arthur's death.

should elicit the response:

> What is Excalibur?

In 2011, IBM had skin in the game. The company had developed a *Jeopardy*-playing computer by the named of Watson. Watson was to compete against the two best *Jeopardy* players in history—Ken Jennings and Brad Rutter. The prize was a cool $1 million.

Ken Jennings held the record for the longest unbeaten winning streak—74 games on the trot. Along the way, he had accumulated

$2.5 million in prize money. Jennings, now 36, had been a computer programmer before his *Jeopardy* success.

Brad Rutter had amassed the greatest winnings in *Jeopardy* history—$3.25 million. Rutter was four years Jennings' junior and worked in a record store before his first appearance on the show.

The seeds of the IBM *Jeopardy* Challenge were sown six years previously. At the time, IBM management were on the lookout for a spectacular computing event. They wanted something that would capture the public imagination and demonstrate the capabilities of IBM's latest machines.

With this in the back of his mind, Director of IBM Research Paul Horn happened to be on a team night out at a local restaurant. Mid-meal the other diners left their tables *en mass* and congregated in the bar. He turned to his colleagues and asked, 'What's going on?'. Horn was told that everyone else was watching *Jeopardy* on TV. Ken Jennings was on his record-breaking winning streak. Half the country wanted to see if he could keep it going. Horn paused and wondered if a computer could play *Jeopardy*.

Back at base, Horn pitched the idea to his team. They didn't like it. Most of the researchers reckoned that a computer wouldn't stand a chance at *Jeopardy*. The topics were too wide ranging. The questions were too cryptic. There were puns, jokes, and double meanings—all stuff that computers weren't good at processing. Regardless, a handful of staff decided to give it a go.

One of the volunteers, Dave Ferrucci, later became Principal Investigator on the project. A graduate of Rensselaer Polytechnic Institute in New York, Ferrucci had joined IBM Research straight out of college after gaining a PhD in Computer Science. His specialism was knowledge representation and reasoning. He was going to need that expertise—this was the toughest natural language processing and reasoning challenge around.

IBM's first prototype was based on the team's latest research. The machine played about as well as a five-year-old child. The Challenge wasn't going to be easy. Twenty-five IBM Research scientists were to spend the next four years building Watson.

By 2009, IBM were confident enough to call the producers of *Jeopardy* and suggest that they put Watson to the test. The show executives arranged a trial against two human players. Watson didn't perform at all well. Its responses were erratic—some were correct, others ridiculous. Watson quickly became the butt of the quizmaster's jokes. The computer was the fool in the room. The machine wasn't ready.

IBM tried again one year later. This time, Watson gained the approval of the TV show producers.

Game on.

The contest was recorded at Watson Center and subsequently broadcast over three consecutive days (14–16 February 2011).

The host is *Jeopardy* regular, Alex Trebek. Bedecked in a grey suit, pink shirt, red tie, and wire frame glasses, Trebek is centre stage—the epitome of calm sophistication. His grey hair is neatly cut, his brown eyes sharp, and alert. His smooth vocal tones command attention. A giant screen to the left of the garish purple and blue set displays the game board. To the right, the players stand behind individual podia emblazoned with their names and prize money totals.

Jennings and Rutter flank a computer monitor displaying an animated graphic. The graphic—a blue cartoon of the world crowned with exclamation marks—is Watson's visible presence—the computer's *avatar*. The machine's robotic thumb rests ominously on the buzzer. Jennings sports a yellow tie, lilac shirt, and dark jacket. His ginger-brown hair is side parted. Rutter's shirt is open necked beneath a black jacket adorned with a stylish pocket square. Rutter's hair is dark brown. His facial hair is somewhere between designer stubble and a full bread. The auditorium is packed with IBM top brass, researchers, and engineers. The highly partisan crowd is charged, noisy, and enthusiastic. This is a home game for Watson.

Rutter picks a category. Trebek reads the first clue of the day. Simultaneously, an equivalent text file is fed to Watson. The clue is: [185]

Four-letter word for a vantage point or a belief.

Rutter is first on the buzzer:

What is a view?

Correct. $200 to Rutter. Trebek: [185]

Four letter word for the iron fitting on the hoof of a horse or a card dealing box in a casino.

Watson is first this time:

What is a shoe?

Correct. $400 to Watson. A television camera picks out Ferrucci's smiling face in the audience.

The clues range from the Beatles to the Olympics. At the end of the first game, the scores are close. Jennings lags on $2,000, and Watson and Rutter are tied on $5,000 apiece.

In game two, Watson starts well but some of its replies are distinctly odd. In response to the clue:

US Cities: Its largest airport is named for a World War II hero, its second largest for a World War II battle.

Watson replies:

What is Toronto?

The correct answer is Chicago. Toronto isn't even in the US.

Nonetheless, Watson wins the game. The final scores for game two are: Watson $35,734, Rutter $10,400, and Jennings a paltry $4,800.

All is still to play for in game three. Early on, Jennings and Watson are neck and neck, out in front of Rutter. Watson gets a clue right and hits a Daily Double. Ferrucci punches the air in delight. Jennings visibly implodes. Later, he says: [185]

That's the moment when I knew it was over.

The final accumulated scores are Watson $77,147, Jennings $24,000, and Rutter $21,600. IBM Watson wins. In the circumstances, IBM donates the $1 million first prize to charity.

After the match, Rutter opines: [185]

I would have thought that technology like this was years away.

For the first time, it seems that conversational artificial intelligence might be within grasp. Jennings offers: [185]

I think we saw something important today.

How did Watson achieve the seemingly impossible?

Certainly, Watson's processing power and prodigious memory were part of the computer's success.

Watson's hardware was bang up-to-date. The machine consisted of a network of one hundred IBM Power 750 servers with a total of fifteen TeraBytes of memory, and 2,880 processor cores. At full tilt, the device could perform eighty trillion calculations per second.

Watson was in possession of a vast trove of data. The rules required that the machine be disconnected from the Internet while the match was in progress. During development, the team had downloaded one million books to Watson. All sorts of important documents were crammed into its memory—textbooks, encyclopaedias, religious texts, plays, novels, and movie scripts.

Despite this, the real secret of Watson's success lay in its algorithms.

Watson's Secret Recipe

Watson's software is an amalgam of hundreds of cooperating algorithms. To begin with, a *parser* algorithm breaks the clue up into its constituent grammatical components. The parser determines what part of speech every word in the clue belongs to. This is done by looking up the words in a dictionary.

Given the clue: [188]

> Poets & Poetry: He was a bank clerk in the Yukon before he published *Songs of a Sourdough* in 1907.

Watson spots that 'he' is a pronoun, 'was' is a verb, and 'bank clerk' is a compound noun.

Based on the identified sentence structure, the parser extracts the relationships between the words. For example, Watson detects a 'was' relationship between 'he' and 'bank clerk', and a 'publish' linkage between 'he' and '*Songs of a Sourdough*'. In addition, it spots an 'in' relationship between 'he' and 'Yukon'.

After finding the explicit relationships between words, Watson hunts for implicit links. The original terms are looked up in a thesaurus to find synonyms. This provides deeper insights into the meaning of the clue. For example, the 'publish' relationship implies an 'author of' link.

Once the relationships have been extracted, the *elements* of the clue are identified. This is done using a set of if-then-else rules applied to the parser output. Three main elements are identified: the clue *focus*, the *answer type*, and the *question classification*. The focus of the clue is the person, event, or thing that the clue is guiding the contestants towards. The answer type is the nature of the focus. The question classification is the category that the clue belongs to. Possible categories include factoid, definition, multiple-choice, puzzle, and abbreviation. In the example, the focus is 'he'—an individual male; the answer type is 'clerk' and 'writer' (implicit); plus the question classification is 'factoid'—a short factual piece of information.

Once clue analysis is complete, Watson searches for answers in its database. Watson launches a number of searches. These searches access the *structured* and *unstructured* data held in Watson's memory banks. Structured data is the name for information held in well-organized tables. Structured tabular data is great for factoid lookup. For example, Watson can look up *Songs of a Sourdough* in a table containing the titles and writers of well-known songs. However, given the obscurity of the poem, this hunt is likely to be fruitless.

Unstructured data is the term for information which is not formally organized. Unstructured data includes information held in textual documents, such as newspapers or books. Plenty of knowledge is contained therein but it is difficult for a computer to interpret. Retrieving useful information from unstructured data turned out to be one of the biggest problems in building Watson. In the end, the team found some surprisingly effective tricks.

One technique involves searching for an encyclopaedia article mentioning all of the words in the clue. Often, the title of the article is the sought-after answer. For example, searching Wikipedia for the words 'bank clerk Yukon Songs of a Sourdough 1907' returns an article entitled 'Robert W. Service'. This is the correct answer.

Another option is to search for a Wikipedia entry whose title is the focus of the clue. The algorithm then hunts for the desired information in the body of the selected article. For example, Watson handles the clue 'Aleksander Kwasniewski became the president of this country in 1995' by looking up an article entitled 'Aleksander Kwasniewski' in Wikipedia. The computer scans the article for the most frequently occurring country name.

Watson launches a battery of such searches in the hope that one will yield accurate results.

The resulting candidate answers are assessed by calculating how well the answers meet the requirements of the clue. Every aspect of the answers and the clue are compared and scored. The answer with the highest score is selected as the best solution. The score is compared with a fixed threshold. If the score exceeds the threshold, Watson rephrases the solution as a question and presses the buzzer. If called on, Watson offers the question to the quizmaster.

Watson's roots lie in the *expert systems* and *case-based reasoning* technologies of the 1970s and 1980s.

Expert systems use hand written if-then-else rules to transform textual inputs into outputs. The first popular expert system, MYCIN, was developed by Edward Feigenbaum's team at Stanford University. It was designed to assist physicians in determining whether an infection is bacterial or viral. Bacterial infections can be treated with antibiotics, whereas viral infections are unresponsive to medication. Doctors commonly over-prescribe antibiotics, mistakenly recommending them for viral infections. MYCIN assists the prescribing physician by asking a series of questions. These probe the patient's symptoms and the outcomes of diagnostic tests. The sequence of questions is determined by lists of hand-crafted rules embedded in the MYCIN software. MYCIN's final diagnosis—bacterial or viral—is based on a set of rules defined by medical experts.

Case-based reasoning (CBR) systems allow for more flexible decision-making than expert systems. The first working CBR system is widely regarded to have been CYRUS, developed by Janet Kolodner at Yale University. CYRUS is a natural language information retrieval system. The system holds the biographies and diaries of US Secretaries of State Cyrus Vance and Edmund Muskie. By referral to these information sources, CYRUS enters into a dialog with the user, answering questions about the two subjects. For example: [190]

Question: Who is Cyrus Vance?
Answer: Secretary of State of the United States.
Q: Does he have any kids?
A: Yes, Five
Q: Where is he now?
A: In Israel

CYRUS generates candidate answers by matching the query with passages from the documents. All matches found are scored according to similarity. The candidate answer with the greatest score is phrased appropriately and returned to the user.

The main drawback with expert systems is that every rule and consideration must be manually programmed into the system. Case-based reasoning, on the other hand, requires that every potential linguistic nuance in the questions and source material is dealt with by program. Due to the complexity of natural language, the mapping between a question and valid answer is complex and convoluted. Algorithms must deal with a bewildering variety of sentence structures. Puns, jokes, and double meanings make the problem even worse. Every subtlety of language presented in the input text, and the reference documents, makes the CBR algorithm more complex.

Watson's success in playing *Jeopardy* hinged on the clues. *Jeopardy* employs a large, but limited, number of question types. With a Herculean effort, the team managed to write algorithms to deal with *Jeopardy*'s most common clue styles. Had the *Jeopardy* producers suddenly changed the format of the clues, Watson would have struggled. In contrast, Watson's human opponents would, mostly likely, have adapted. Watson was programmed to deal with *Jeopardy* clues and nothing else. Despite the appearance of genius, Watson understands nothing. It simply shuffles words according to predefined rules. It was just that Watson had more rules and data than any previous natural language processing system.

Yet, IBM didn't build Watson just for playing *Jeopardy*: [185]

[Watson is] about doing research in deep analytics and in natural language understanding. This is about taking the technology and applying it to solve problems that people really care about.

After the Challenge was over, IBM set up a business unit to commercially exploit the technologies developed in building Watson. That business unit is now focused on healthcare applications, especially automated diagnosis of medical conditions. Retargeting of Watson's highly customized algorithms has proven tricky, though. Even IBM executives have admitted that progress has been slower than expected. [193]

With hindsight, the version of Watson that competed in the *Jeopardy* Challenge was the pinnacle of case-based reasoning. Much more powerful AI technologies were waiting in the wings. Hints of the coming revolution were embedded in Watson. Here and there, a handful of tiny artificial neural networks augmented Watson's decision-making capabilities. These networks were portents of the future. The dinosaurs of AI—expert systems and CBR—were about to be swept away by a tsunami.

11

Mimicking the Brain

The machinery of recall is thus the same as the machinery of association, and the machinery of association, as we know, is nothing but the elementary law of habit in the nerve-centres.

WILLIAM JAMES
The Principles of Psychology, 1890 [194]

Humans have an innate ability to recognize patterns. In just a few formative years, children learn to recognize faces, objects, voices, smells, textures, and the spoken word. Throughout the twentieth century, researchers failed miserably in their attempts to design algorithms that could match human prowess in pattern recognition. Why were computers so good at arithmetic, yet so poor at pattern recognition?

To better understand the conundrum, imagine developing a system to recognize cats in photographs.

The first step is to convert an image to an array of numbers that a computer can process. The lens of a digital camera focuses light onto a grid of electronic sensors. Each sensor converts the light level to a number in the range 0 to 1. In a *greyscale* image, 0 indicates black, 1 white, and the values in-between shades of grey. Every number corresponds to a dot, or *pixel* (picture element), in the image. The array of numbers is a digital approximation of the image. This much is straightforward. The challenge in pattern recognition is not in creating a digital image, but in writing an algorithm capable of making sense of the numbers.

The difficulty arises from the variability of real-world images. First, there are lots of breeds of cat. The cat could be fat or thin, big or small, furry or hairless, grey, brown, white, or black, tailed or not. Second, the cat could be in any one of a great many poses—it might be lying down, sitting, standing, walking, or jumping. It could be looking at the camera, to the left, to the right, or have its back to the lens. Third, the photograph could have been taken under a range of conditions.

It might be daylight or night, flash photography, a close-up, or a long-shot. Writing an algorithm to cope with each and every circumstance is extremely difficult. Every possibility requires a new rule which has to interact with all of the old rules. Pretty soon, the rules begin to conflict. Eventually, algorithm development grinds to a halt.

Rather than write millions of rules, a handful of computer scientists pursued an alternative line of enquiry. Their contention was simple. If the best pattern recognition engine in the world is the human brain, why not just replicate it?

Brain Cells

By the onset of the twentieth century, neuroscientists had established a basic understanding of the human nervous system. The work was spearheaded by Spanish neuroscientist Santiago Ramón y Cajal, who received the Nobel Prize for Medicine in 1906.

The human brain is composed of around 100 billion cells, or *neurons*. A single neuron is made up of three structures: a central body, a set of input fibres called *dendrites*, and a number of output fibres called *axons*. When inspected under a microscope, the long thin wispy dendrites and axons stretch away from the bulbous central body, branching as they go. Every axon (output) is connected to the dendrite (input) of another neuron via a tiny gap, called a *synapse*. Neurons in the brain are massively interconnected. A single neuron can be connected to as many as 1,500 other neurons.

The brain operates by means of electrochemical pulses sent from one neuron to another. When a neuron *fires*, it sends a pulse from its central body to all of its axons. This pulse is transferred to the dendrites of the connected neurons. The pulse serves to either *excite* or *inhibit* the receiving neurons. Pulses received on certain dendrites cause excitation, and pulses received on others cause inhibition. If a cell receives sufficient excitation from one or more other neurons, it will fire. The firing of one cell can lead to a cascade of firing neurons. Conversely, pulses arriving at inhibitory neural inputs reduce the level of excitation, making it less likely that a neuron will fire. The level of excitation, or inhibition, is influenced by the frequency of the incoming pulses and the sensitivity of the receiving dendrite.

Canadian neuropsychologist Donald Hebb discovered that when a neuron persistently fires, a change takes place in the receiving dendrites.

Figure 11.1 Walter Pitts. (*Courtesy Science Photo Library.*)

They become more sensitive to the firing neuron. As a consequence, the receiving neuron becomes more likely to fire in response. Hebb's discovery revealed a *learning effect* in biological neural networks whereby past experience determines future action. The finding offered a crucial link between the workings of individual neurons and the higher-level learning capability of the brain.

In 1943, two American researchers suggested that the workings of neurons could be modelled by means of mathematics. The proposal was as radical as it was intriguing.

Walter Pitts (Figure 11.1) was a child genius. Coming from a disadvantaged background in Detroit, Pitts taught himself mathematics, Greek, and Latin from the books in the public library. Browsing the aisles one day, he spotted *Principia Mathematica*. Far from an easy read, the tome establishes the logical foundations of mathematics. In reading the book, Pitts espied a number of flaws. He wrote a letter to Bertrand Russell, one of the authors, pointing out the mistakes. Delighted, Russell replied, inviting Pitts to study with him at Cambridge University. Unfortunately, Pitts couldn't go. He was twelve years old.

At fifteen, Pitts heard that Russell was scheduled to speak at the University of Chicago. Pitts ran away from home, never to return. He snuck into Russell's lectures and subsequently took a menial job at the college. Homeless, Pitts happened to befriend Jerome Lettvin, a young medical student. Lettvin saw something special in Pitts and introduced him to Warren McCulloch.

McCulloch, a Professor at the University of Illinois, was Pitts' senior by twenty-four years. In stark contrast to Pitts' childhood, McCulloch was raised in a well-to-do, professional, east coast family. He studied psychology before obtaining a degree in Neurophysiology. The unlikely pair, McCulloch and Pitts, entered into deep conversations about McCulloch's research. At the time, McCulloch was attempting to represent the function of neurons by means of logical operations. Pitts grasped McCulloch's intent and suggested an alternative, mathematical approach to the problem. Seeing the young man's potential and predicament, McCulloch invited both Pitts and Lettvin to live with him and his wife.

Ensconced in this new abode, Pitts worked late into the night with McCulloch. They developed the idea that the state of a neuron could be represented by numbers. Furthermore, they postulated that the firing patterns of interconnected neurons could be simulated by means of equations. They developed mathematical models demonstrating that networks of neurons could perform logical operations. They even showed that such networks could assume some of the functions of a Turing machine (see Chapter 3).

Aided by McCulloch, Pitts procured a graduate position at MIT, despite the fact that he had not finished high school. Professor of Mathematics Norbert Wiener, one of his mentors at MIT, guided Pitts towards *cybernetics*, the study of self-regulating systems. The subject encompasses all kinds of self-regulating systems, from biology to machines. Perhaps the most common examples are today's thermostatically controlled heating systems. Pitts, Weiner, McCulloch, and Lettvin formed a loose affiliation to further the field. They reached out to like-minded researchers such as John von Neumann. Later, Lettvin said of the cyberneticians: [199]

> [Pitts] was in no uncertain terms the genius of our group. He was absolutely incomparable in the scholarship of chemistry, physics, of everything you could talk about—history, botany, etc. When you asked him a question, you would get back a whole textbook. To him, the world was connected in a very complex and wonderful fashion.

In 1952, McCulloch was invited by Wiener to lead a neuroscience project at MIT. He jumped at the chance to work with Pitts once again and promptly moved to Boston. Lamentably, Pitts began to struggle with bouts of melancholy coupled with alcohol addiction.[200] Abruptly and without explanation, Weiner cut off all contact with Pitts, McCulloch, and Lettvin. Floundering, Pitts slipped into depression. He died from an alcohol-related illness in 1969 at just 46 years of age. McCulloch passed away four months later from a heart condition, aged 70. Their legacy was the mathematical foundation for the artificial neural network.

Artificial Neural Networks

The world's first artificial neural network (ANN) was built by Belmont Farley and Wesley Clark at MIT in 1954. The pair constructed a simple simulation of firing neurons in a computer program. They used numbers to represent the state of the neurons and track the sensitivities of the inputs and outputs. The network was programmed to recognize images of simple binary numbers (sequences of ones and zeros). While Farley and Clark were first, it was Frank Rosenblatt (Figure 11.2) who popularized the ANN concept.

Rosenblatt was from New Rochelle, New York (born 1928). He studied at Cornell University, obtaining a PhD degree in Psychology before taking up a faculty position at the university. To begin with, he simulated an ANN on an IBM computer (1957). Later, to increase the speed

Figure 11.2 Frank Rosenblatt, inventor of the Mark 1 Perceptron, 1950. (*Deceased Alumni files, #41-2-877. Division of Rare and Manuscript Collections, Cornell University Library.*)

of processing, he built an ANN in the form of an electronic device: the Mark I Perceptron (1958). Rosenblatt's Perceptron was designed to recognize simple shapes in images. The images (just 20x20 pixels) were fed from a black and white camera into the Perceptron. The Perceptron simulated the behaviour of a small network of neurons. The network output comprised of a set of neuron outputs, each corresponding to one of the shapes to be recognized. The highest output value indicated which shape had been spotted in the image.

Although the Perceptron was very limited, Rosenblatt was a persuasive communicator. After one of his press conferences, the *New York Times* wrote: [205]

> The Navy revealed the embryo of an electronic computer today that it expects will be able to walk, talk, see, write, reproduce itself and be conscious of its existence. Later Perceptrons will be able to recognize people and call out their names and instantly translate speech in one language to speech and writing in another language, it was predicted.

The gap between Rosenblatt's predictions and reality was immense. Nevertheless, his demonstrations, papers, and book were highly influential in spreading the concept of ANNs to other research institutes.

A Perceptron is a *classifier*—it determines which *class*, or category, a given input belongs to. In Rosenberg's case, the input to the Perceptron was an image of a shape. The goal of the Perceptron was to determine whether the shape was a circle, a triangle, or a square. These shapes were the recognized classes.

A Perceptron consists of one or more layers of artificial neurons. Each neuron has a number of inputs and a single output (Figure 11.3). The strength of the signal on an input or output is represented by a number. The output from a neuron is calculated based on its inputs. Each input is multiplied by a *weight*. The weight models the sensitivity of the neuron to signals on that particular input. The weighted inputs are summed together with a *bias* value to give the neuron's *excitation*. The excitation is applied to an *activation function*. In a Perceptron, the activation function is a simple threshold operation. If the excitation is greater than the threshold (a fixed numeric value), then the output is 1. If it is below the threshold, the output is 0. The threshold value is fixed for the entire network. The 0 or 1 arising from the activation function is the final output of the neuron.

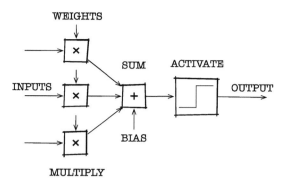

Figure 11.3 Artificial neuron.

The weights and biases of the neurons are collectively called the *parameters* of the network. Their values determine the conditions under which the neurons fire, that is, output a 1. Parameters can be positive or negative numbers. An input of 1 applied to a positive weight increases the excitation of the neuron, making it more likely to fire. Conversely, an input of one multiplied by a negative weight decreases the neuron's excitation, making it less likely to fire. The bias determines how much input excitation is needed before the neuron exceeds the fixed threshold and fires. During normal network operation, the parameters are fixed.

Perceptrons take a number of inputs (Figure 11.4). For example, Rosenblatt's 20x20 pixel image is input to the Perceptron on 400 connections. Each connection has a 0 or 1 value depending whether the associated pixel is black or white. The input connections are fed into the first layer of the network: the input layer. Neurons in the first layer only have a single input connection. Thereafter, the outputs from one layer feed into the inputs of the next layer. In a *fully connected* network all outputs from one layer are input to every neuron in the following layer. The output of the input layer is connected to the first *hidden* layer. Hidden layers are those not directly connected to either the network inputs or outputs. In a simple network, there might be only one hidden layer. After the hidden layers comes the output layer. The outputs from the neurons in this layer are the final network outputs. Each output neuron corresponds to a class. Ideally, only the neuron associated with the recognized class should fire (i.e. output a 1).

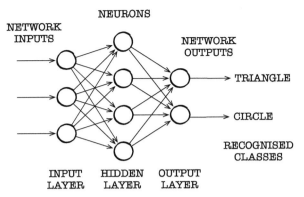

Figure 11.4 Perceptron with three inputs, two fully connected layers, and two output classes.

In summary, an ANN simulation proceeds as follows:

Take the network input values.
Repeat for every layer:
  Repeat for every neuron in the layer:
    Set the total excitation equal to the bias.
    Repeat for every input to the neuron:
      Multiply the input value by the input weight.
      Add to the total excitation.
    Stop repeating when all the inputs have been processed.
    If the excitation is greater than the threshold,
    then set the neuron output to 1,
    else set the neuron output to 0.
  Stop repeating when the layer has been processed.
Stop repeating when all layers have been processed.
Output the class name associated with the greatest network
  output.

The overall operation of an ANN can be visualized by imagining the neurons lighting up when they fire. An input is applied to the network. This drives 1s and 0s into the input layer. Some of the neurons fire, sending 1s into the hidden layer. This excites some of the hidden neurons and they fire as a result. This stimulates one of the neurons in the output layer. This single output neuron fires, indicating which pattern was observed at the network input.

Every neuron makes a tiny decision about the input. Working in cooperation, a large number of interconnected neurons can make complex decisions about the nature of the input. The capacity to recognize complex patterns arises from the orchestration of all these tiny decisions.

There are two challenges in ANN design. The first is selecting a suitable *topology*. This is the arrangement of neurons in the network. The topology determines the number of layers, the quantity of neurons in each, and their interconnection. The topology affects the complexity of the task that the network can handle. Generally speaking, additional neurons and layers are needed for more complex pattern recognition tasks. The second challenge is determining the values for the parameters of the network. The parameters control the behaviour of the network. If the network is to correctly classify the inputs, the parameter values must be just right.

Rosenblatt chose his network topology based on experience coupled with trial-and-error, something that hasn't really changed to this day.

For a given topology, Rosenblatt employed a training procedure to find effective parameter values. Training started with random parameters. Rosenblatt then fed a few example inputs into the network and checked to see if the output was correct. If it wasn't correct, he tweaked the parameters until the Perceptron gave the correct answer. He repeated this process for a set of input examples. Once training was finished, Rosenblatt assessed the accuracy of his Perceptron by testing it with previously unseen inputs. Rosenblatt referred to his training method as *back-propagating error correction*. More casually, it might be said that he twiddled the knobs on his Perceptron until it worked. The method was laborious but gave reasonably good results in recognizing a small number of simple shapes.

Perceptrons ran into criticism and controversy in the late 1960s. Marvin Minsky and Seymour Papert published a book entitled *Perceptrons* that poured cold water on the whole idea. Their book carried weight in the research community. Both authors were well regarded MIT Professors and Directors of MIT's Artificial Intelligence Laboratory. Papert, a South African, was a mathematician by training. Before arriving at MIT, he had accumulated an impressive CV, including sojourns at the University of Cambridge, the University of Paris, the University of Geneva, and the NPL. Minsky, from New York City, had originally been a neural network believer. His PhD thesis at Princeton was on that very topic. He even designed an electronic device with learning capability inspired by brain synapses: the SNARC (1952). Thereafter, Minsky became a convert to the school of *symbolic logic* (see Chapter 5). For him, neural networks were out, and logical reasoning was in.

Their book provided a mathematical analysis of the properties of Perceptrons. It described the strengths of Perceptrons but highlighted two important limitations. First, they pointed out that single-layer Perceptrons cannot perform certain elementary logical operations. Second, they argued that Rosenblatt's training method would not work for multilayer Perceptrons. The book was scathing:[206]

> Perceptrons have been widely publicized as 'pattern recognition' or 'learning' machines and as such have been discussed in a large number of books, journal articles, and voluminous 'reports'. Most of this writing [...] is without scientific value. Many of the theorems show that Perceptrons cannot recognize certain kinds of patterns.

The proponents of Perceptrons bristled. Minsky and Papert's two central criticisms were valid but misleading, they claimed. True, single-layer Perceptrons cannot learn certain primitive logic functions, but multi-layer Perceptrons *can*. True, Rosenblatt's training method does not work for multi-layer Perceptrons but that does not mean that a suitable training procedure cannot be found. In a review of the book, H.D. Block retorted:[208]

> A Perceptron, as defined by Minsky and Papert, is slightly more general than what Rosenblatt called a simple Perceptron.
>
> On the other hand, the simple Perceptron [...] is not at all what a Perceptron enthusiast would consider a typical Perceptron. He would be more interested in Perceptrons with several layers, feedback, and cross coupling. In summary then, Minsky and Papert use the word Perceptron to denote a restricted subset of the general class of Perceptrons.

Some observers concluded that Minsky and Papert were intentionally trying to kill the Perceptron. Whatever their rationale, the book contributed to a decline in Perceptron research. Amidst growing disenchantment with artificial intelligence, the field as a whole was starved of funds and the first AI Winter set in.

Tragically, Rosenblatt died in 1971 on his 43rd birthday in a sailing accident in Chesapeake Bay. Minsky and Papert dedicated the second edition of *Perceptrons* to his memory.

In the years that followed, Minsky remained a prolific and celebrated AI researcher and writer. He won the ACM Turing Award in 1969 for his contributions to advancement of the field. Minsky and Papert passed away in 2016, both aged 88.

How to Train a Brain

After 1970, only a handful of computer scientists persevered with ANNs. The big problem was finding a way to train multilayer networks. In the end, the problem was solved four times over by four independent research teams. Communication was so poor that no one knew what anyone else was doing. It would be 1985 before word got out that multilayer ANNs could be trained by means of an algorithm.

At Harvard, Paul Werbos solved the problem as part of his PhD. His algorithm—*back-propagation* or *backprop*—had been around for a while but had not previously been applied to ANN training. Werbos told Minsky—by then the doyen of AI—about his solution:[214]

> In the early 1970s, I did in fact visit Minsky at MIT. I proposed that we do a joint paper showing that [multilayer Perceptrons] can in fact overcome the earlier problems. But Minsky was not interested.

More than a decade later (1985), David Parker described backprop in an MIT technical report. That same year, French student Yann LeCun published a paper describing an equivalent method at a conference in Paris. The year after, a letter describing backprop appeared in the prestigious journal *Nature*. The authors were David Rumelhart and Ronald Williams (University of California, San Diego), and Geoffrey Hinton (Carnegie Mellon University). Unaware of the previous publications, the group had been working on the idea for a few years. The letter clearly set out the backprop algorithm and its application to ANNs. With the imprimatur of *Nature* behind it, backprop was finally established as the algorithm for training ANNs.

Backprop requires a minor change to the activation function of artificial neurons. The threshold operation must be replaced by a *smoother* function. The new function ensures that the neuron output rises gradually from 0 to 1 as the excitation increases. Gone is the sudden thresholded transition from 0 to 1 used in the Perceptron. The smooth 0 to 1 transition allows the network parameters to be progressively adjusted during backprop.

This smoother activation function also means that the network outputs have a range of possible values between 0 and 1. Thus, the final decision is no longer the output connection with a 1 value. Instead, it is the output connection with the largest value. The change has the side effect of improving robustness when the input is on the cusp of two classes.

Normal operation of an ANN is called *forward-propagation* (or *inference*). The ANN accepts an input, processes the values neuron-by-neuron, layer-by-layer, and produces an output. During forward-propagation, the parameters are fixed.

Backprop is only used in training, and operates within a machine learning framework (see Chapter 9). To begin, a dataset containing a large number of example inputs and associated outputs is assembled. This dataset is split into three. A large training set is used to determine the best parameter values. A smaller validation set is put aside to assess performance and guide training. A test set is dedicated to measuring the accuracy of the network after training is finished.

Training begins with random network parameters. It proceeds by feeding an input from the training set into the network. The network processes this input (forward-propagation) using the current parameter vales to produce an output. This network output is compared with the desired output for that particular input. The error between the actual and desired output is measured as the average difference between the actual and desired outputs squared.

Imagine that the network has two classification outputs: circle and triangle. If the input image contains a circle then the circle output should be 1, and the triangle output 0. Early in the training process, the network probably won't work at all well, since the parameters are random. So, the circle output might have a value of $\frac{2}{3}$ and the triangle $\frac{1}{3}$. The error is equal to the average of $(1 - \frac{2}{3})$ squared and $(0 - \frac{1}{3})$ squared, that is $\frac{2}{9}$.

The parameters in the network are then updated based on the error. The procedure commences with the first weight in the first neuron of the output layer. The mathematical relationship between this particular weight and the error is determined, and this relationship is used to calculate how much the weight should change to reduce the error to zero. This result is reduced by a constant value, called the *learning rate*, and subtracted from the current weight. The weight adjustment has the effect of reducing the error if the network is presented with the same input again. Multiplication by the learning rate ensures that the adjustment is gradual and averaged over a large number of examples. These steps are repeated for all of the parameters in the network, moving backwards through the layers.

Error calculation, backprop, and parameter update are performed for every input–output pair in the training dataset. Over many iterations, the error is gradually reduced. In effect, the network learns the relationships between the input examples and the desired output classes. Training ends when no further reduction in error is observed (a summary of the algorithm is provided in the Appendix).

An ANN's great strength lies in its ability to learn and *generalize*. That is, the network determines the correct class for inputs that it has never seen before, but which are similar to those that it was trained on. In other words, a network trained with many drawings of circles, will correctly classify a sketch of a circle that it has never seen before. The neural network doesn't just memorize the training data—it learns the general relationship between the inputs and the output classes.

Backprop enabled researchers to efficiently train multilayer networks for the first time. As a result, networks became more accurate and capable of more complex classification tasks. By the end of the decade, it was proven, at least in theory, that a sufficiently large multilayer network could learn any input-output mapping. Minsky and Papert's objections had been overruled.

Be that as it may, the benefits of ANNs were not obvious to most observers. Even with backprop, networks were still small. The computers of the day just weren't up to performing the vast number of calculations needed to train a large network. ANNs were a mere sideshow for the next twenty years (1986–2006). Sure, they might be a useful aid to cognitive scientists attempting to comprehend the workings of the brain, but they weren't something for serious computer scientists and electronic engineers. Proper classification algorithms relied on rigorous mathematics and statistics, not hocus-pocus.

Amidst widespread scepticism, a few success stories hinted at the potential of ANNs. One of the highlights of the period was a piece of work involving Yann LeCun (Figure 11.5), the same Yann LeCun who, as a PhD student, had presented backprop at an obscure conference in Paris in 1985.

Figure 11.5 Artificial neural network innovator Yann LeCun, 2016. (*Courtesy Facebook.*)

Recognizing Digits

LeCun was born in Paris in 1960. He received the Diplôme d'Ingénieur degree from the École Supérieure d'Ingénieurs en Electrotechnique et Electronique (ESIEE) in 1983. In his second year, he happened upon a philosophy book discussing the nature versus nurture debate in childhood language development. Seymour Papert was one of the contributors. From there, he found out about Perceptrons. He started reading everything he could find on the topic. Pretty soon, LeCun was hooked. He specialized in neural networks for his PhD at Université Pierre et Marie Curie (1987). After graduation, LeCun worked for a year as a post-doctoral researcher in Geoffrey Hinton's lab at the University of Toronto in Canada. By then, Hinton was an established figure in the neural network community, having been a co-author on the backprop letter published in *Nature*. A year later, LeCun moved to AT&T Bell Laboratories in New Jersey to work on neural networks for image processing.

LeCun joined a team that had been working on building a neural network to recognize hand written digits. Conventional algorithms weren't at all good at this—there was too much variability in the writing styles. Casually written 7s are easily confused with 1s, and vice versa. Incomplete 0s can be interpreted as 6s, and 2s with long tails are often mixed up with truncated 3s. Rule-based algorithms just couldn't cope.

The team acquired a large dataset of digital images by scanning the zip codes from the addresses on envelopes passing through the Buffalo, New York, post office. Every letter yielded five digits. In the end, the dataset incorporated 9,298 images. These examples were manually sorted into ten classes corresponding to the ten decimal digits (0 to 9).

The team developed an ANN to perform the recognition task but had little success with it. The complex mapping necessitated a large network. Even with backprop, training the network had proven difficult. To solve the problem LeCun suggested an idea that he had tinkered with in Hinton's lab.

The ANN took a 16x16 pixel greyscale image of a single digit as input. The network output consisted of ten connections—one for each digit class between 0 and 9. The output with the strongest signal indicated the digit recognized.

LeCun's idea was to simplify the network by breaking it up into lots of small networks with shared parameters. His approach was to create a unit containing just twenty-five neurons and a small number of layers. The input to the unit is a small portion of the image—a square of 5x5 pixels (Figure 11.6). The unit is replicated sixty-four times to create a *group*. The units in the group are tiled across the image, so

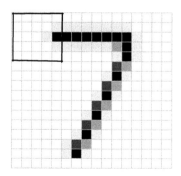

Figure 11.6 Grayscale image of the digit 7 (16×16 pixels). The pixel inputs to the unit at the top left are highlighted. Sixty-four copies of this unit are spread over the image.

that it is entirely covered in units. Every unit's input overlaps with its neighbour's by three pixels.

The overall network contains twelve groups. Since the units in a group share the same parameters, all perform the same function but are applied to a different part of the image. Each group is trained to detect a different feature. One group might detect horizontal lines in the picture, another vertical, still another diagonal. The outputs from each group are fed into a fully connected three-layer network. These final layers fuse the information coming from the groups and allow recognition of the digit in its entirety.

The network is hierarchical in structure. A single unit detects a 5×5 motif in the image. A group spots a single motif anywhere in the image. The twelve groups detect twelve different motifs across the image. The final, fully connected layers detect the spatial relationships between the twelve motifs. This hierarchical organization draws inspiration from the human visual cortex, wherein units are replicated and successive layers process larger portions of the image.

The beauty of LeCun's scheme is that all of the units in a single group share the same weights. As a consequence, training is greatly simplified. Training the first layers in the network only involves updating twelve units, each containing just twenty-five neurons.

The mathematical process of replicating and shifting a single unit of computation across an image is called *convolution*. Hence, this type of network became known as a *convolutional neural network*.

The Bell Labs convolutional neural network proved to be extremely effective, achieving a breathtaking accuracy of ninety-five per cent. The network was close to human accuracy. The team's findings were published in 1989 and the system commercialized by AT&T Bells Labs. It was estimated that, in the late 1990s, ten to twenty per cent of bank cheques signed in the US were automatically read by a convolutional neural network.

In 2003, Yann LeCun left Bell Labs to be appointed Professor of Computer Science at New York University. Meanwhile, Geoffrey Hinton, his old mentor at the University of Toronto, was putting together a posse.

Deep Learning

Hinton (Figure 11.7) was born in post-war Wimbledon, England (1947). Hinton reckons that he wasn't particularly good at maths in school. Nevertheless, he gained entry to the University of Cambridge, enrolling in Physics and Physiology. Dissatisfied, he transferred to Philosophy.

Figure 11.7 Deep neural network pioneer Geoffrey Hinton, 2011. (*Courtesy Geoffery Hinton.*)

Finally, he settled on Psychology. Looking back, Hinton says that he wanted to understand how the human mind works. He concluded that the philosophers and psychologists didn't have the answers. He turned to computer science.

On graduation, the young man worked as a carpenter for a year before embarking on a PhD at Edinburgh University. With the grudging acquiescence of his supervisor, Hinton persisted in pursing research on ANNs. On completion of his doctorate, Hinton wended the itinerant path of the fledgling academic. He worked at the University of Sussex, the University of California-San Diego, Carnegie Mellon University, and University College London (UCL) before joining the University of Toronto as a Professor.

In 2004, Hinton submitted a proposal requesting funding for a research project focused on neural computation to the Canadian Institute For Advanced Research (CIFAR). CIFAR was known for funding basic research, but it was still a long shot. Yoshua Bengio, Professor at the Université de Montréal, later commented:[223]

> It was the worst possible time. Everyone else was doing something different. Somehow, Geoff convinced them.

The modest grant paid for a series of invite-only meetups involving some of the top ANN researchers in the world. Bengio again:[223]

> We were outcast a little bit in the broader machine learning community: we couldn't get our papers published. This gave us a place where we could exchange ideas.

The grant turned out to be the beginnings of a tectonic shift.

In 2006, Hinton, Simon Osindero, and Yee-Whye Teh (University of Toronto, National University of Singapore) published a transformative paper. The document marked the beginnings of what is now known as *deep learning*. The paper described a network made up of three fully connected hidden layers. The network had so many parameters that training by means of backprop was prohibitively slow. To solve the problem, Hinton and team devised a novel procedure to accelerate training.

Normally, backprop starts with random parameter values. Instead, the team inserted a pretraining stage before backprop. The purpose of this new stage was to quickly find a good set of parameters from which backprop could start.

Backprop is an example of *supervised* training. This means that the network is provided with examples of matching inputs and outputs. In this new preliminary stage, Hinton and his co-authors proposed employing *unsupervised* training. Unsupervised training only uses input examples.

During unsupervised pre-training, example inputs are fed into the network. The network parameters are adjusted by an algorithm so that the ANN learns to detect significant patterns in the input. The network isn't told what classes these patterns are associated with—it just learns to distinguish the patterns. For handwriting recognition, these patterns might be the length and direction of lines or the position and length of curves. To achieve this, the training algorithm updates the parameters for just one layer at a time, starting with the input layer. In other words, the algorithm grows the network parameters from the input forward. This approach has significantly lower computational complexity than backprop.

Once pretraining is complete, the network is able to distinguish the most prominent patterns in the input dataset. After that, supervised training is applied as normal, commencing with the pretrained parameters. Since backprop has a good starting point, it requires far fewer iterations to complete training.

Following in the footsteps of Bell Labs, Hinton's team elected to tackle hand written digit recognition. This time, a much larger dataset was available. The project used the MNIST dataset amassed by LeCun, Corinna Cortes (Google Labs), and Christopher Burges (Microsoft Research). MNIST contains 70,000 hand written digits culled from US census returns and high school exam scripts.

The resulting ANN achieved an accuracy of 89.75 per cent, which was not as good as LeCun's convolutional neural network. However, that wasn't the point. They had proven that, by means of pretraining, a deep, fully connected network could be trained. The road to deeper and more effective networks was open.

Over the course of the next decade, deep learning gained momentum. The confluence of three advances enabled researchers to build larger and deeper networks. Smarter algorithms reduced computational complexity, faster computers reduced run-times, and larger datasets allowed more parameters to be tuned.

In 2010, a team of researchers in Switzerland conducted an experiment to see if increasing the depth of a neural network really did translate into improved accuracy. Led by long-time neural network guru Jürgen Schmidhuber, the group trained a six-layer neural network to recognize digits. Their network contained a whopping 5,710 neurons. They, like Hinton's group, used the MNIST dataset of hand written digits. However, even MNIST wasn't big enough for Schmidhuber's team's purposes. They artificially generated additional digit images by distorting the MNIST photographs.

The resulting ANN achieved an accuracy of 99.65 per cent. This wasn't just a world record, this was human-level performance.

Suddenly, it dawned on everyone that ANNs had been too small to be of any practical use. Deep networks were the way to go. A revolution in artificial intelligence was at hand.

The Tsunami

The deep learning tsunami hit in three waves: first, speech recognition, then image recognition, next natural language processing. Half a century of pattern recognition research was swept away in just three years.

For sixty years, the tech community had struggled to accurately convert spoken words to text. The best algorithms relied on the Fourier transform (see Chapter 2) to extract the amplitude of the harmonics. Hidden Markov Models (HMMs) were then used to determine the phonemes uttered based on the observed harmonic content and the known probability of sound sequences in real speech.

With the help of Navdeep Jaitly, an intern from Hinton's Lab, Google ripped out half of their production speech recognition system and replaced it with a deep neural network. The resulting hybrid ANN–HMM speech recognition system contained a four-layer ANN. The team trained the ANN with 5,870 hours of recorded speech sourced from Google Voice Search, augmented with 1,400 hours of dialogue from YouTube. The new ANN–HMM hybrid outperformed Google's old HMM-based speech recognition system by 4.7 per cent. In the context of automatic speech recognition, this was a colossal advance. With his mission at Google accomplished, Jaitly—intern extraordinaire—returned to Toronto to finish his PhD.

Over the course of the next five years, Google progressively extended and improved their ANN-based speech recognition system. By 2017, Google's speech recognition system had attained ninety-five per cent accuracy—a previously unheard-of level accuracy.

In 2012, Hinton's group reported on a deep neural network designed to recognize real-world objects in still images. The objects were everyday items such as cats, dogs, people, faces, cars, and plants. The problem was a far cry from merely recognizing digits. Digits are made up of lines, but object identification requires analysis of shape, colour, texture, and edges. On top of that, the number of object classes to be recognized greatly exceeded the paltry ten Hindu–Arabic digits.

The network—dubbed AlexNet after lead designer Alex Krizhevsky—contained 650,000 neurons and sixty million parameters. It incorporated five convolutional layers followed by three fully connected layers. In addition, the work introduced a simple, yet surprisingly effective, technique. During training, a handful of neurons are selected at random and silenced. In other words, they are prevented from firing. *Drop-out*, as the technique was named, forces the network to spread the decision-making load over more neurons. This has the effect of making the network more robust to variations in the input.

The team entered the network into the ImageNet Large Scale Visual Recognition Challenge in 2012. The dataset for the competition consisted of approximately 1.2 million training images and 1,000 object classes. Krizhevsky, Ilya Sutskever, and Hinton's deep convolutional network swept the boards. AlexNet achieved a top five accuracy of 84.7 per cent. That is to say, the true object class was among the ANN's top five picks more than 84 per cent of the time. The network's error rate was almost half that of the second placed system.

Meanwhile, just 500 km east along the St. Lawrence River from Toronto, a team at the Université de Montréal was investigating how deep neural networks could be applied to the processing of text. That team was led by Yoshua Bengio (Figure 11.8).

Hailing from Paris, France (born 1964), Bengio was one of the leading lights of the neural network renaissance. He studied Electronic Engineering and Computer Science at McGill University in Montreal, obtaining BEng, MSc, and PhD degrees. A science fiction fan as an adolescent, Bengio became passionate about neural network research as a graduate student. He devoured all of the early papers on the topic. A self-professed nerd, he set out to build his own ANN. After working as a post-doctoral researcher at AT&T Bell Labs and MIT, Bengio joined the Université de Montréal as a faculty member in 1993. Bengio's team trained ANNs to predict the probability of word sequences in text.

In 2014, Google picked up on Bengio's work and adapted it to the problem of translating documents from one language to another. By then, the Google Translate web service had been in operation for eight years. The system relied on conventional approaches to segment sentences and map phrases from one language to another. On the whole,

Figure 11.8 Neural network researcher Yoshua Bengio, 2017. (© *École polytechnique - J. Barande.*)

the system's translations weren't particularly good. The translated sentences were mostly readable but hardly fluent.

Google took the unusual step of connecting two neural networks back to back. In the scheme, the output of the first network—the *encoder*—feeds into the output of the second—the *decoder*. The idea was that the encoder would convert text written in English to an abstract vector of numbers. The decoder would then reverse the process, converting the abstract vector of numbers into French. The researchers didn't specify the intermediate number vector. They simply relied on the training procedure to find a suitable representation.

After two years of effort, Google completed development of an eight-layer encoder and a matching eight-layer decoder. The network was trained on a corpus of thirty-six million manually translated sentence pairs. The new system outperformed the previous Google Translate production system, reducing the number of translation errors by an impressive sixty per sent. When the system went live on the Google web site, bilingual users reported a sudden and dramatic improvement in translation quality.

Success after success bred a deep learning stampede. Companies foresaw a plethora of new applications powered by deep learning— self-driving cars, smart cameras, next-generation recommenders, enhanced web search, accurate protein structure prediction, expedited drug design, and many more. Google, Facebook, IBM, Apple, Amazon, Yahoo!, Twitter, Adobe, and Baidu snapped up deep learning talent. Rumours abounded of seven-figure starting salaries for the neural network rock stars. LeCun was appointed Director of AI Research at Facebook. Andrew Ng joined Baidu as Chief Scientist. At the age of 65, Geoffrey Hinton became a Google summer intern!

In 2015, amidst the gold rush, LeCun, Hinton, and Bengio published a paper in *Nature* surveying developments. Deep neural networks had swept the entire field of artificial intelligence before it. Everything was changed, utterly.

LeCun, Hinton, and Bengio were the recipients of the 2018 ACM Turing Award, sharing the Google sponsored $1 million prize.

With the runaway success of deep learning, some have speculated that human-level artificial general intelligence (see Chapter 5) is just around the corner. LeCun demures: [236]

Whether we'll be able to use new methods to create human-level intelligence, well, there's probably another fifty mountains to climb, including ones we can't even see yet. We've only climbed the first mountain. Maybe the second.

All we have to date are sophisticated pattern recognition engines. Yet, we may speculate on the path that might take us across these mountains. Presently, the best guess is that a network of ANNs will be required. Significant improvements may also require a fundamental reworking of the ANN. Today's ANNs are only a rough approximation of what goes on in a biological neural network. It may be that a more realistic model is needed. The devil may well be in the details.

For those outside the computer science community, the first hint of the power of deep neural networks came in 2016. In that year, an artificial intelligence hit the world's news media headlines. Albeit in a narrow field of endeavour, an artificial intelligence had, perhaps for the first time, attained superhuman ability.

12

Superhuman Intelligence

Because previous moves cannot be changed, subsequent regrets
are truly difficult to bear.

Unknown author, translated by Robert W. Foster
The Classic of Go, 6th century [237]

19 March 2016. A young man walks purposefully down the corridors of
Seoul's Four Seasons Hotel. Along the way, he passes ranks of journalists
and photographers clamouring for his attention. Dressed in a sharp
navy suit and open-necked shirt, he looks much younger than his
thirty-three years. The man is thin and Asian in appearance. His hair
is combed from the crown of his head towards an unwavering fringe.
His upper lip bears the ghost of a moustache. Despite all the attention,
he seems relaxed and confident.

The man leaves the cacophony of the corridors behind and enters a
hushed conference room. A small audience and a handful of television
cameras face a low, neon blue stage. The man settles into a black leather
chair set on the right of a low podium. The lettering on the podium
indicates that this is 'Lee Sedol'. A South Korean flag confirms his
nationality.

Opposite Lee is Aja Huang. The lettering on Huang's side of the
podium spells out one word: 'AlphaGo'. Although Huang is Taiwanese,
the flag below is British. A computer monitor, keyboard, and mouse
rest alongside Huang. A panel of judges sits behind the two men,
overlooking them. A table separates Lee and Huang. The table bears a Go
board, a timer, and four bowls. Two of the bowls are empty. One holds
a collection of identical white stones, the other similar black pieces.

Lee is widely acknowledged to be one of the top five Go players in
the world. He has won eighteen international titles. A child prodigy,
he studied at the famous Korean Baduk (Go) Association. Lee turned

professional at just twelve years of age. Among Go aficionados, he has a reputation for aggressive and imaginative play. Lee Sedol—the 'Strong Stone'—is a national figure in South Korea.

AlphaGo is a computer program. The codification of a complex game-playing algorithm. Huang's job is to relay Lee's moves to AlphaGo and to place stones on the board for the computer. AlphaGo is the creation of a small London-based outfit by the name of DeepMind Technologies. Huang is lead programmer on the AlphaGo project. Two years previously, DeepMind was acquired by Google for an estimated $500–625 million.

Google is the sponsor of the AlphaGo–Lee match. The prize money is set at $1 million. If the computer wins, the money goes to charity.

In the days leading up to the match, Lee was confident. At one press conference, he claimed that the question wasn't whether he would win the match. Rather, it was whether he would lose a single game. Lee seemed justified in his confidence. A computer had never beaten a top 300 professional Go player in a competitive match. Prior to the showdown in Seoul, grandmasters predicted that Lee would pocket an easy million bucks.

Like Chess, Go is an abstract war simulation. The game originated in China around three thousand years ago and spread to Korea and Japan during the fifth to seventh centuries CE. Go remains hugely popular in east Asia.

The Championship version of the game is played on a 19 × 19 grid. At the start, the grid is empty. Players take turns to place one stone on a grid intersection. An intersection is called a *territory*. A player can pass their turn, if they so wish. One player places black stones, the other white. The aims of the game are to enclose territory and capture enemy stones. A player's stones are removed from the board when they are encircled by their opponent's pebbles. The game ends when both players pass consecutively rather than play moves. A player can resign by placing a stone off-grid. The player with the greatest number of territories plus captives wins. The player that is second to start is given a small points bonus in compensation. In competitive matches, moves are made on a timer.

Watching a video of Go on fast-forward video is hypnotic. Complex patterns of black and white stones evolve, coalesce, and colonize the board in a slow dance. Sudden bursts of activity transform the board as stones are encircled and vanish. Go devotees see an underlying beauty in the game. To them, a match is a reflection of a player's imagination, courage, and fortitude. The values of the game—elegance and humility—are instilled in players from an early age.

Although the rules are simple to learn, Go is highly complex to play. A Go board is more than five times the size of a Chess board (8x8). Go games last, on average, 150 moves. In making a move, a Go player must consider around 250 possibilities. The theoretical Go game tree (see Chapter 5) contains an astronomical 250^{150}, or 10^{359}, nodes. By this estimate, Go is 10^{226} times (a one with 226 zeros after it) more complex than Chess.

The Match

The AlphaGo-Lee match is a best-of-five games contest. An estimated sixty million television viewers are watching the game on television in China alone. A hundred thousand enthusiasts are glued to the live English language coverage on YouTube.

The DeepMind team spectates from a war room in the bowels of the hotel. The room is kitted out with a wall of monitors. Some screens display camera feeds from the match room. Others show lists of numbers and graphs summarizing AlphaGo's analysis of the game. DeepMind CEO Demis Hassabis and lead project researcher Don Silver watch the match unfold from this vantage point. Like the rest of their team, Hassabis and Silver are anxious, but powerless.

Day one, game one. Lee places the first stone. Bizarrely, it takes AlphaGo half a minute to respond. The AlphaGo team holds its breath. Is the machine working at all? Finally, it makes its decision and Huang places AlphaGo's first stone.

AlphaGo attacks from the outset. Lee seems mildly surprised. AlphaGo isn't playing like a computer at all. Then comes AlphaGo's move 102. It is aggressive—a gateway to complicated skirmishes. Lee recoils, rubbing the back of his neck. He looks worried. He strengthens his resolve and rejoins battle. Eighty-four moves later, Lee resigns. The reaction in the DeepMind team room is euphoric.

Afterwards, Lee and a composed Hassabis face the assembled media at the post-game press conference. The two sit apart on stools on a bare stage. Lee looks isolated, lost, abandoned. He is deeply disappointed but accepts his loss with grace. The following morning, AlphaGo's victory is front page news.

Day two, game two. This time, Lee knows what to expect. He plays more cautiously. On move 37, AlphaGo makes an unexpected play—a move that humans seldom play. In shock, Lee walks out of the conference room. Huang and the match judges stay put, bewildered. Minutes later, having collected his thoughts, Lee returns to the fray. After 211 moves, Lee again resigns.

AlphaGo's move 37 was decisive. The computer estimated that the chances of a human playing the move was one in ten thousand. The European Go champion Fan Hui was awestruck. For him, move 37 was, 'So beautiful. So beautiful.' AlphaGo had displayed insight beyond human expertise. The machine was creative.

At the press conference, Lee reflected on the game:[242]

> Yesterday, I was surprised. But today I am speechless. If you look at the way the game was played, I admit, it was a very clear loss on my part. From the very beginning of the game, there was not a moment in time when I felt that I was leading.

Day three, game three. Lee's facial expressions say it all—initial calm, turning to concern, followed by agony, and finally dismay. He resigns after four hours of play. Against all expectations—save those of Google and DeepMind—AlphaGo wins the match.

Lee looks worn out. Regardless, he is gracious in defeat:[243]

> I apologize for being unable to satisfy a lot of people's expectations. I kind of felt powerless.

A strange kind of melancholy descends on proceedings. Everyone is affected, even the DeepMind team. Those present are witnessing the suffering of a great man. One of Lee's rivals remarks that Lee had fought:[238]

> A very lonely battle against an invisible opponent.

Even though the series is decided, Lee and AlphaGo press on, playing games four and five. In game four, Lee is more himself. The Strong Stone takes a high-risk strategy. His move 78—a so-called 'wedge' play—is later referred to by commentators as 'God's move'. AlphaGo's response is disastrous for the machine. Soon its play becomes rudderless. Finding no way out, the computer begins to make nonsense moves. Eventually, AlphaGo resigns.

Lee tries the same high-risk approach in game five. This time, there is no miracle play. Lee is forced to resign.

AlphaGo wins the match by 4 games to 1.

The Winning Move

AlphaGo's victory sent shockwaves through both the Go and the computer science communities. Based on projections of computer performance, this wasn't supposed to happen for at least another fifteen years. The theory was that tackling Go needed far faster hardware than was available in 2016. In reality, the secret to AlphaGo's success lay in its algorithms, not in its hardware.

The AlphaGo hardware was mundane by the standards of 2016. During development, the DeepMind team used only forty-eight central and eight graphics processing units—something that a hobbyist could easily rig together in their garage. In competition, AlphaGo ran on computers in one of Google's Internet-connected data centres. The program occupied 1,920 central and 280 graphics processing units. In contrast, the most powerful supercomputer at the time—the Chinese Tiamhe-2—had 3.1 million central processing units. AlphaGo was a slouch by comparison.

Like Arthur Samuel's Checkers playing program, the AlphaGo algorithm utilizes Monte Carlo tree search (see Chapter 5). On the computer's turn, it hunts for the most promising next play. For each of these, it examines the mostly likely responses by its opponent. Following that, it evaluates its own possible replies. In this way, the computer produces a tree of possible future plays with the current board position as its root.

Once the tree has been built, the computer uses a minimax procedure to select the best move (see Chapter 5). The computer starts with the furthest look-ahead board positions—the leaves of the tree. It then moves backwards through the tree towards the root. At every branching point, it propagates the best move backwards in the tree. On its own plays, the best choice is the one that maximizes the computer's chances of a win. On its opponent's plays, the computer selects the play that minimizes its own probability of winning. When the procedure reaches the root of the tree, the computer selects the play which, it considers, will give it the best chance of winning the game in the long term.

AlphaGo uses ANNs to evaluate board positions. A board position is represented by a table of numbers. Each number indicates whether there is a black stone, a white stone, or no stone at a grid intersect. To evaluate a position, the table of numbers is input to an ANN. The neural network outputs a score indicating the strength of the position.

AlphaGo's neural networks are larger versions of the convolutional neural networks introduced by Yann LeCun to recognize digits (see Chapter 11). In effect, the table of digits is treated much like an image. AlphaGo's neural networks recognize the patterns on the board, much as LeCun's network recognized the lines and curves in numerals.

AlphaGo uses three neural networks.

The first is a *value network*. The value network estimates the probability of winning from a given position. The value network scores the positions at the end of the tree search.

The second ANN is a *policy network*. The policy network serves to guide the tree search. The policy network scores a position based on how promising it is. If a position looks like it might lead to a win in the future, it is given a high policy score. Only positions with high policy scores are investigated in greater depth. In this way, the policy network controls the breadth of the search.

If the value network was completely accurate then tree search would not be necessary. The computer could simply evaluate all of the next positions and choose the best. The lookahead improves accuracy by rolling the positions forward. As the game gets closer to the end, it becomes easier to predict the outcome and so the value network becomes more accurate.

Ideally, the value network would also be used for policy decisions. Again, the value network isn't sufficiently accurate. A separate policy network offers greater accuracy in evaluating early game positions. The value network is trained for precision, whereas the policy network is trained not to miss promising paths in the tree search.

The third network is a SL-value (Supervised Learning) network. This network is trained to score positions in the same way as humans. The other networks seek to determine the true chances of winning. The SL-value network allows the computer to predict the most likely move that a human player will make.

AlphaGo's neural networks were trained in three stages.

In stage one, the SL-value network was trained using supervised learning (see Chapter 11). The network contained thirteen layers of neurons. Training was performed using positions and moves obtained from the KGS Go database. KGS allows players from all over the world to play Go online for free. Games are recorded and are available on the KGS web site. AlphaGo used this database to supply examples of board positions and the moves played by humans. Thirty million positions from 160,000 games were used to train the network.

In stage two, the SL-value network was refined to create a policy network. This time, *reinforcement learning* was used. The network played Go against itself. The reinforcement algorithm used the outcome of every game (win or lose) as a reference with which to update the network parameters. AlphaGo played 1.2 million games against a pool of older versions of itself. As it played more games, AlphaGo's performance gradually edged upwards. In trials, the resulting policy network beat the original SL-value network in eighty per cent of games.

In stage three, the team used the policy network to seed a value network. Again, reinforcement learning was used. Rather than playing entire games from scratch, intermediate positions from the KGS database were used as starting points. A further thirty million games were played to complete training of the value network.

These three neural networks were the chief difference between AlphaGo and prior Go playing computers. Previously, positions were evaluated using hand-crafted rules and scoring methods (i.e. expert systems and case-based reasoning). AlphaGo's ANNs offered far more accurate position evaluation.

The accuracy of AlphaGo's ANNs comes from the confluence of three factors. First, deep ANN are extremely good at learning complex relationships between inputs and outputs. Second, during training, the networks were exposed to huge volumes of data. In preparation for the match, AlphaGo inspected more Go moves than any human, ever. Third, advances in algorithms and hardware allowed these large networks to be trained in a reasonable time frame.

These factors explain why AlphaGo outperformed previous programs, but they do not explain why AlphaGo beat Lee Sedol. Analysis of Go and Chess grandmasters thought processes suggest that they evaluate far fewer positions than AlphaGo. The human tree search is much narrower and shallower than the computer's. Human pattern recognition must, therefore, be much more effective than AlphaGo's. AlphaGo makes up for this deficiency by means of faster processing. In competition, the machine evaluates many more board positions than humans. The high speed of AlphaGo's electronic simulations of neuron behaviour allowed it examine more positions in the tree search. That is how AlphaGo beat Lee Sedol.

The implication was that there remained plenty of scope to improve the pattern recognition capabilities of ANNs.

DeepMind

To outsiders, it must have seemed that DeepMind was an overnight success but, of course, it wasn't. Demis Hassabis, the company's co-founder and CEO, had been thinking about board games and computers since he was a kid.

Hassabis (Figure 12.1) was born in London, England, in 1976. He is proudly 'North London born and bred'.[245] Hassabis reached master level in Chess aged 13. He spent his winnings on his first computer—a Sinclair Spectrum 48K—and taught himself to program. Before long, he had completed his first Chess-playing program.

Hassabis finished high school at sixteen and joined a video game development company (Lionhead Studios). A year later, he was co-designer and lead-programmer on the popular management simulation game *Theme Park*. Hassabis left the company to enrol in a Computer

Figure 12.1 DeepMind co-founder and CEO Demis Hassabis, 2018. (*Courtesy DeepMind.*)

Science degree programme at Cambridge University. In his spare time, he entered the Pentamind competition at the annual Mind Sports Olympiad. Pentamind pits elite players against each other across five board games: Backgammon, Chess, Scrabble, Go, and Poker. Hassabis went on to win the competition a record five times.

Reflecting on the range of his accomplishments, Hassabis says:[246]

> I get bored quite easily, and the world is so interesting, there are so many cool things to do. If I was a physical sportsman, I'd have wanted to be a decathlete.

After graduation from Cambridge, Hassabis founded his own independent video games development company. Elixir Studios released two games but then hit problems. The firm was wound up in 2005. Hassabis's formal statement on the closure belied his disappointment at the turn of events:[247]

> It seems that today's games industry no longer has room for small independent developers wanting to work on innovative and original ideas.

Hassabis resolved that his career should take a new direction. He set out on a mission to build artificial intelligence. Believing that the best first step lay in understanding how biological intelligence works, he embarked on a PhD in Cognitive Neuroscience at UCL. The subject explores how the human brain works, often employing computer models to better comprehend brain function. Hassabis published a string of important research papers on the topic before graduating.

Armed with these fresh insights, Hassabis co-founded DeepMind Technologies with Shane Legg and Mustafa Suleyman in 2010. Hassabis and Suleyman had known another since childhood. Hassabis and Legg met while studying for their PhDs at UCL .

DeepMind first came to the attention of the wider scientific community thanks to a letter published in *Nature* magazine. The letter described an artificial neural network that DeepMind trained to play Atari video games. Atari video games are the coin-operated classics from the arcades of the 1980s, including Space Invaders, Breakout, and River Raid. DeepMind's neural network took the screen image as input. Its output was the control signals for the game—the joystick twitches and the button presses. In effect, the ANN replaced the human player.

DeepMind's ANN taught itself how to play Space Invaders from scratch. Its sole preprogrammed aim was to score as many points as possible. At first, the network played randomly. Through trial and error and a learning algorithm, it gradually accumulated a suite of point-scoring tactics. By the end of training, DeepMind's neural network was better at Space Invaders than any previous algorithm. This, in itself, was an achievement. What was remarkable was that the network went on to learn how to play forty-nine different Atari video games. The games were varied, requiring different skills. Not only could the network play the games, it could play them just as well as a professional human games tester. This was new. DeepMind's ANN had excelled across a range of tasks. For the first time, an ANN was showing a general-purpose learning capability.

A year later—and just two months before the Lee match—DeepMind published another paper in *Nature*. In it, they described AlphaGo and casually mentioned that the program had beaten the European Go champion Fan Hui. The paper should have been a warning to Lee Sedol and others. However, Europe was regarded as a Go backwater. It was presumed that Fan Hui had erred. Fan Hui, for his part, was so impressed with AlphaGo that he accepted an offer to act as a consultant to the DeepMind team as they prepared for the Lee Sedol match.

AlphaGo's resounding victory over Lee garnered headlines world-wide. In contrast, AlphaGo's later defeat of the world number one was an anticlimax. AlphaGo beat nineteen-year-old Ke Jie 3–0 in May 2017. This time, the match received little media coverage. The world seemed to have accepted humankind's defeat and moved on. After the win, Hassabis said that, based on AlphaGo's analysis, Jie had played almost perfectly. Almost perfect was no longer good enough. After the match, DeepMind retired AlphaGo from competitive play.

Yet the company didn't stop working on computers that played Go. It published another paper in *Nature*, describing a new neural network program dubbed AlphaGo Zero. AlphaGo Zero employed a reduced tree search and just one neural network. This single two-headed network replaced the policy and value networks of its predecessor. AlphaGo Zero used a new, more efficient training procedure based exclusively on reinforcement learning. Gone was the need for a database of human moves. AlphaGo Zero taught itself how to play Go from scratch in a meagre forty days. In that time, it played twenty-nine million games. The machine was allowed just five seconds of processing time between moves. AlphaGo Zero was tested against the version of AlphaGo that beat Ke Jie. AlphaGo Zero won 100 games to nil.

In just forty days the computer had taught itself to play Go better than any human, ever. AlphaGo Zero was emphatically superhuman.

Human Go grandmasters pored over AlphaGo Zero's moves. They discovered that AlphaGo Zero employed previously unknown game-winning strategies. Ke Jie began to include the new tactics in his own repertoire. A new age in the history of Go was dawning. Human grandmasters were now apprentices to the machine. Biological neural networks were learning from their artificial creations.

The true significance of AlphaGo Zero doesn't lie on a Go board, however. Its real importance lies in the fact that AlphaGo Zero is a prototype for a general-purpose problem solver. The algorithms embedded in its software can be applied to other problems. This capability will allow ANNs to rapidly take on new tasks and solve problems that they haven't seen before—something that heretofore only humans and high-level mammals have achieved.

The first signs of this general problem-solving capacity appeared in 2018 in yet another *Nature* paper. This time, the DeepMind team trained an ANN named AlphaZero to play Go, Chess, and Shogi (Japanese Chess). It wasn't particularly surprising to read that AlphaZero learned how to play the three games solely from self-play. Neither was it especially eyebrow raising to note that AlphaZero defeated the previous world-champion programs (Stockfish, Elmo, AlphaGo Zero) at all three games. What was jaw dropping was that, starting from random play, AlphaZero learned to play Chess in just nine hours, Shogi in twelve hours, and Go in thirteen days. The human mind was beginning to appear weak in comparison.

13

Next Steps

In terms of cryptocurrencies, generally, I can say with almost certainty that they will come to a bad ending.

WARREN BUFFET
On CNBC, 2018[251]

Computer algorithms have fundamentally altered the way that we live. Information technology is deeply embedded in our workplaces. Communication is entrusted to email, social media, and messaging apps. Our leisure time is dominated by video games, streaming music, and online movies. Recommenders manipulate our purchasing decisions. Our romantic liaisons are prompted by algorithms. Much has changed. Yet, many more revolutionary technologies are under development in the hip open-plan offices of the tech giants, the makeshift workspaces of impoverished start-ups, and the tatty labs of college professors. In this, the final chapter, we examine two new algorithms that have the potential to change the world.

Cryptocurrency

The first of these is the algorithm that underpins *cryptocurrency*. Cryptocurrency is a form of money that only exists as information held in a computer network. The world's first cryptocurrency, Bitcoin, now has over seventeen million 'coins' in circulation with a total real-world value of $200 billion (2019). Cryptocurrencies seem set to disrupt the global financial system.

The origins of cryptocurrency lie in the Cypherpunk movement that began in the 1990s. The Cypherpunks are a loose amalgam of skilled cryptographers, mathematicians, programmers, and hackers who believe passionately in the need for electronic privacy. Connected by mailing lists and online discussion groups, the Cypherpunks develop open-source software that users can deploy for free to secure their data and communications. Their ideals were set out by Eric Hughes in *A Cypherpunk's Manifesto* (1993):[253]

> Privacy is necessary for an open society in the electronic age.
>
> We cannot expect governments, corporations, or other large, faceless organizations to grant us privacy out of their beneficence.
>
> Privacy in an open society [...] requires cryptography.
>
> Cypherpunks write code. We know that someone has to write software to defend privacy, and since we can't get privacy unless we all do, we're going to write it. We publish our code so that our fellow Cypherpunks may practice and play with it. Our code is free for all to use, worldwide.

The Cypherpunks contributed their skills to a series of projects to develop secure software. PGP enabled RSA encryption for email users. Tor allowed anonymous web browsing. The group wrote white papers on matters of encryption. They filed lawsuits against the US government in regard to export controls on encryption technologies. On occasion, they even urged public disobedience in support of their aims. The Cypherpunks also promoted the concept of cryptocurrency.

In their eyes, cryptocurrency has three key advantages over conventional currency. Firstly, a cryptocurrency is not controlled by a central authority. There is no central bank of Bitcoin. The currency is managed by a network of computers. Anyone can join the network. There is no application form. Volunteers simply download the cryptocurrency software from the Internet and run it. No computer on the network is more important than any other. All are peers. Secondly, users are anonymous, provided that they do not trade *cryptocoins* for conventional currency. Privacy is guaranteed by means of public key cryptography. Anyone can be a user. They simply download an app, which submits their transactions to the network. Thirdly, transactions carry a low fee and zero sales tax. Furthermore cryptocoins can be sent internationally without incurring currency exchange charges.

While the Cypherpunks were early proponents of cryptocurrency, no one knew how to make it work for real. There didn't seem to be any way around the Double-Spend Problem.

Conventional online currencies rely on a central authority that approves transactions (i.e. a transfer of money between users). The central authority maintains a ledger of all transactions. The ledger is the electronic equivalent of the pen-and-paper logs maintained by banks 100 years ago. By inspection of the ledger, the central authority knows how much money every user has in their account. When a user requests a transaction, the central authority can easily check if they have enough money to cover the transaction. If they do, the transaction is accepted as valid and recorded in the ledger. If they don't, the transaction is rejected.

The difficulty in designing a cryptocurrency is the removal of the central authority. The ideal is that a distributed network of computers maintains the ledger. Every computer on the network has its own copy of the ledger. The hard part is synchronizing these copies of the ledger (i.e. keeping them all up to date). Communication delays are highly unpredictable on the Internet. Computers can join and leave the network at any time. These issues lead to the Double-Spend Problem.

Imagine that Alice only has 1.5 cryptocoins in her account. She owes money to both Bob and Charlie. In desperation, she sends two transactions to the network. In one, she transfers 1.5 cryptocoins to Bob. In the other, she transfers 1.5 cryptocoins to Charlie. If she sends the two transactions to different parts of the network, there is a chance that one computer will accept the transfer to Bob at exactly the same time as another device accepts the transfer to Charlie. If she is lucky, Alice will pay both parties at the same time, *double-spending* her funds.

Bitcoin

Satoshi Nakamoto announced a solution to the Double-Spend Problem on 31 October 2008 in a white paper posted to a Cypherpunk mailing list. The paper introduced Bitcoin, the world's first practical cryptocurrency. The following January, Nakamoto released the Bitcoin source code and the original—or *genesis*—Bitcoin block.

Fundamentally, bitcoins are just sequences of characters (numbers and letters) held in a computer network. A bitcoin only has value because people believe that it has value. Users expect that they will be able to exchange bitcoins for goods and services at a later date. In this, Bitcoin is no different from the banknotes in your pocket. The paper itself has little intrinsic value. Its value derives from the expectation that you will be able to exchange it for something of worth.

Bitcoin is reasonably straightforward to use. Users buy, sell, and exchange bitcoins via apps. Bitcoins can be used to purchase real-world goods from participating retailers. Digital brokers will happily exchange bitcoins for old-fashioned state-controlled currency. User anonymity is protected by public key cryptography (see Chapter 7). Prior to using Bitcoin, a user generates a public and private key pair. They keep the private key secret. The public key functions as the user's ID on Bitcoin.

When a user wants to send bitcoins to another user, they create a transaction. The transaction consists of the transaction ID, the sender's ID, the recipient's ID, the amount, and the input transaction's IDs (Figure 13.1). The inputs to the transaction are previous transactions in which the sender received the bitcoins that they are about to spend. The input transactions must already have been logged in the ledger and must not have been spent previously. Say Alice wants to send ฿0.5 to Bob. She references two previous transactions in which she received ฿0.3 from Jack and ฿0.2 from Jill. She does this by including the IDs of the previous transactions in the new transaction. The amount to be spent and the total amount referenced must match exactly. This may mean that the sender has to send some bitcoins back to themselves as change.

The sender's encryption key is used to *authenticate* a transaction (Figure 13.2). Authentication ensures that the sender really did want to transfer that bitcoin amount to the receiver. It also guarantees that the transaction isn't from a duplicitous third party. To enable authentication, the sender appends a digital signature to the transaction. A digital signature is the equivalent of the handwritten signature on a paper cheque.

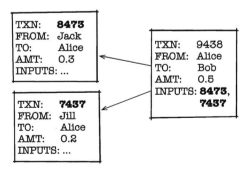

Figure 13.1 A Bitcoin transaction must refer to previous input transactions with the same total value.

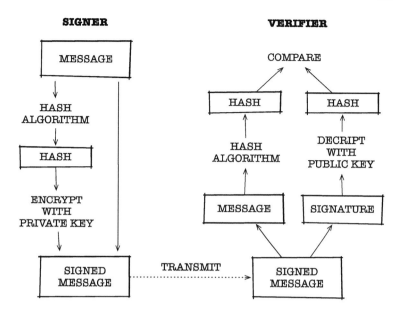

Figure 13.2 Creation and verification of a digital signature.

The digital signature is created by encrypting a summary of the transaction with the sender's private key. Normally, a public key is used for encryption and a private key for decryption. To generate a signature, the process is reversed. The private key encrypts and the public key decrypts. This means that anyone can check the signature but only the sender can create it.

When a computer on the Bitcoin network receives the transaction, it firstly verifies that the signature is genuine (Figure 13.2). It does this by decrypting the signature using the sender's public key. This gives the transaction summary. The receiver also summarises the transaction and compares the two versions—the decrypted and the calculated. If they match, the transaction must be authentic. Only the real sender could have created the digital signature since only they have the private key. If the two versions do not match, the transaction is rejected as being invalid.

The digital signature ensures that the transaction wasn't tampered with en route. Any change to the message alters its summary. As a result, the calculated summary will not match the decrypted equivalent.

The transaction summary is calculated by means of a *hashing* algorithm. A hashing algorithm takes a large amount of text and squeezes it down to a shorter sequence of characters. Information is lost in the process, but the output sequence is highly dependent on the input. In other words, a small alteration in the original text leads to a large random change in the output. The summary is often called the *hash* of the message. The hashing function used in Bitcoin is essentially an advanced checksum algorithm (see Chapter 7).

After authentication, the new transaction is *validated*. The receiving computers check that the input transactions exist in the ledger and that the associated monies have not been spent previously.

Blockchain

Next, the transaction is *confirmed* and *logged*. The network computers incorporate the new transaction in a larger block of unconfirmed transactions. A block is simply a group of unconfirmed transactions and their associated data. The network computers race to add their blocks to the ledger. The winner of the race incorporates its block in the chain of blocks. This chain of blocks—or *Blockchain*—is the ledger (Figure 13.3). It links every confirmed Bitcoin block in an unbroken sequence stretching all the way back to Nakamoto's genesis block. The links of the chain are formed by including the ID of the previous block in the next block. The chain rigidly defines the order in which transactions are applied to the ledger. The transactions in a single block are considered to have happened at the same time. Transactions in any previous block are

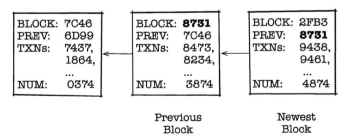

                                           Previous   Newest
                                             Block   Block

Figure 13.3 The Blockchain. Blocks are chained together by block ID to enforce the order in which the transactions are applied to the ledger. Blocks contain a unique block ID, the previous block's ID, a group of transactions, and a randomly generated number.

considered to have happened at an earlier time. The winner of the race shares its block with the entire network so that all of the ledgers are kept up to date.

The race ensures that only one computer at a time can add a block to the chain. Winning the race depends on luck. In most cases, there is a significant delay between the first and second placed computers. This delay allows time for the Blockchain update from the winner to propagate through the network. Dead heats between first and second are unlikely, but can occur. Provision is made for this by requiring that six blocks are added to the chain before a block is considered confirmed. Six dead heats in succession are, for practical purposes, impossible.

To win the race, a computer has to create a valid block. This is done by generating a random number and appending it to a candidate group of transactions. A hashing algorithm is then applied to the block. If the hash is less than a pre-determined threshold, the candidate block is considered valid. If the hash is equal to, or greater than, the threshold, the candidate block is viewed as invalid. The network computers try different random numbers until a valid block is formed. The first computer to create a valid block wins the race and shares the block with the rest of the network. The other machines log the transactions by adding the block to their own copy of the ledger. All of the machines then resume their attempts to create a new valid block using transactions which have not yet been committed to the ledger.

Creating a valid block amounts to trying lots of random numbers. It is impossible to predict in advance which number will pass the hash-threshold test. Trial and error is the only way to find a suitable number. Since producing a valid block is down to luck, any computer might be the next winner. Thus, there is no single authority for adding blocks to the chain. The work is distributed across the entire network of peer computers (see the Appendix for further details).

This explains how bitcoins are exchanged between users. But how are they created in the first place?

Every time that a network computer creates a valid block, its owner is rewarded with bitcoins. The process of validating blocks in return for bitcoins is called *mining*. Validation of the genesis block released fifty bitcoins to Nakamoto. In the same way, the owners of network computers are rewarded for maintaining the ledger.

Bitcoin's early adopters were buyers and sellers in the online black marketplaces. Bitcoin's emergence on the Darknet was fuelled by illegal drug purchases. At first, the main attraction was anonymity. In time, legitimate organizations began to accept Bitcoin payments. Coinbase was established in 2012 as a broker for digital currencies. In 2014, Microsoft began to accept bitcoins for online purchases of Xbox games. The value of a bitcoin in 'hard' currencies became a rollercoaster. In 2011, a single bitcoin (₿1) was worth 30 US cents. On 18 December 2017, the price spiked to a staggering all-time high of $19,498.63. At first glance, mining bitcoins appears to be a way of creating money from nothing. All that is required is to install the Bitcoin software, download the ledger, and start mining. However, the costs of the computer and the electricity that it consumes are real. Estimates put the annual global revenue from Bitcoin mining at over $8 billion with costs of more than $3 billion (2019). The rewards for mining halve every four years. Ultimately, around twenty-one million bitcoins will be issued.

Bitcoin's success has led to a slew of cryptocurrencies, most notably Ethereum (2015) and Facebook's Libra (2019). However, companies are also investing heavily in the Blockchain technology that underpins Bitcoin. Blockchain offers a secure distributed ledger, independent of the cryptocurrency aspect. A Blockchain can track and enforce sequencing on any form of transaction. Possibilities include validating legal contracts, maintaining online identities, recording medical histories, verifying media providence, and tracing supply chains. Blockchain will likely reshape operational models in a wide range of industries and may, ultimately, prove to be more useful that Bitcoin itself.

Who Is Nakamoto?

The really curious thing about Bitcoin is that no one knows who Satoshi Nakamoto—Bitcoin's inventor—is. The first mention of Nakamoto was the release of the original Bitcoin white paper. Nakamoto remained active on the Cypherpunk mailing lists for a few years, then in 2010 Nakamoto passed control of the Bitcoin source code to Gavin Andresen. The following April, Nakamoto declared: [260]

> I've moved on to other things.
> It's in good hands with Gavin and everyone.

Save for a handful of messages—most of which are now thought to be hoaxes—that was the last anyone heard from Nakamoto.

Of course, there has been much speculation regarding Nakamoto's identity. The clues are scant. Nakamoto is clearly a world-class cryptographer (or group of cryptographers). The Bitcoin source code is impeccable. So Nakamoto is also an expert coder. Nakamoto's written English is perfect. Thus it may be that Nakamoto is a native speaker. Closer inspection of Nakamoto's posts reveals hints of a British or Australian accent. The genesis block includes a headline from the London *Times*. Perhaps Nakamoto is from the UK. Analysis of timestamps shows that Nakamoto mostly posted bulletin board messages between 3 pm and 3 am Greenwich Mean Time. If this habit was due to a nocturnal sleeping pattern, then this means that Nakamoto was likely living on the east coast of America. There are no clues as to Nakamoto's gender.

Lists of suspects have been bandied around online. The names of most top-notch Cypherpunks have been suggested at one time or another. Some individuals have even claimed to be Nakamoto. To date, however, no one has proven that they are Nakamoto. All that has to be done to settle the case is to decrypt a message sent using Nakamoto's public key. The person, or persons, that can do that must be in possession of Nakamoto's private encryption key.

Nakamoto's 1.1 million bitcoins remain untouched. Presently, 1.1 million bitcoins are worth in excess of $11 billion (2019). By that reckoning, Nakamoto is one of the richest 150 people in the world. Why doesn't he, she, or they step forward to claim their rightful riches? Is their reluctance simply down to strict observation of the Cypherpunk code of honour? Or is something more sinister going on?

Satoshi Nakamoto remains an enigma.

Quantum Computers

Bitcoin draws heavily on RSA public key cryptography to ensure user anonymity and provide transaction authentication. In turn, the security of the RSA algorithm hinges on the assumption that there is no fast algorithm for prime factorization of large numbers (see Chapter 7). In other words, there is no fast method for determining which two prime numbers were multiplied to produce a given large number. Clearly, the prime factors of 21 are 3 and 7, but this determination is only quick because 21 is small. Prime factorization of a large prime can take decades on a supercomputer.

Bitcoin, and the whole edifice of Internet security, depends on this single assumption of slow prime factorization. If a fast prime factorization algorithm were to be invented, Bitcoin and almost every secret message on the Internet would suddenly become vulnerable to attack. In 1994, the spectre of just such an algorithm came into view. The only saving grace was that the miracle algorithm required a new type of computer. A device called a *quantum computer*.

In 1981, Richard Feynman delivered the keynote address at a conference at MIT. By then, Feynman was sixty-three and widely regarded to be one of the greatest physicists of all time. During the war, he worked on the Manhattan Project at Los Alamos. At Cornell University, he made huge strides in quantum electrodynamics. While at Caltech, Feynman introduced new concepts in superfluidity and quantum gravity. He was one of the recipients of the Nobel Prize for Physics in 1965.

Feynman's talk at MIT was entitled *Simulating Physics with Computers*. In it, he argued that conventional computers will never be up to the job of accurately simulating the behaviour of subatomic particles. He proposed that a new kind of computer was needed. A computer that would use quantum effects to simulate physical systems. His idea was that the weird behaviour of subatomic particles could be exploited to perform calculations at incredibly high speeds. Feynman dubbed his theoretical machine a quantum computer.

For more than a decade, Feynman's idea was an intellectual curiosity—something that mathematicians and physicists toyed with, but no one took seriously. Actually, building such a machine would be an inordinately complex exercise. Plus, there didn't seem to be much point in doing it. Conventional electronic computers were good enough for most tasks.

Then, in 1994, Peter Shor, a Professor of Applied Mathematics at MIT, changed mainstream thinking on quantum computers. He unveiled an algorithm that could perform fast prime factorization on a quantum computer. If a quantum computer could be built, Shor's algorithm would be orders of magnitude faster than any previous method.

Conventional computers represent information by means of voltage levels on microscopic wires. If the voltage level on a wire is high, then the wire represents a one. In contrast, a low voltage level indicates a zero value. This two-level system is called binary since every wire can only have one of two values—0 or 1 (see Chapter 7). Crucially, at any moment in time, the voltage level on each wire has a single value and so represents a single binary digit (or bit). Therefore, calculations have to be performed one after another.

In contrast, quantum computers represent information using the properties of subatomic, or quantum, particles. Various physical properties of subparticles can be used. One option is the spin of an electron. An upward spin might represent a 1, downward a 0. The big advantage of using the properties of subatomic particles is that, in the quantum world, particles can exist in multiple states at the same time. This strange behaviour is encapsulated in the *principle of superposition*. The effect was uncovered by physicists in the early part of the twentieth century. An electron can spin with all possible orientations simultaneously. Exploiting this effect to represent data means that a single electron can represent 0 and 1 simultaneously. This phenomena gives rise to the basic unit of information in a quantum computer—the quantum bit, or *qubit*.

Quantum computers become exponentially more powerful as qubits are added. A single qubit can represent two values—0 and 1—simultaneously. Two qubits allow four values—00, 01, 10, and 11—to be represented at the same time. A ten-qubit system can capture all decimal values from 0 to 1,023, inclusive, simultaneously. When a quantum computer performs an operation, it is applied to all states at the same time. For example, adding one to a ten-qubit system performs 1,024 additions at once. On a conventional computer, these 1,024 additions would have to be performed one after another. This effect bestows on quantum computers the potential for exponential acceleration in calculation.

There is a snag, though. Measuring the value of a qubit *collapses* its state. This means that the qubit settles to a single value when its physical state is measured. Thus, even though a ten-qubit system can perform 1,024 additions simultaneously, only one result can be retrieved. Worse again, the output retrieved is selected at random from the 1,024 possibilities. The collapsed result of the addition of 1 could be any value from 1 to 1,024. Clearly, random selection of the output is not desirable. Mostly, we want to input some data and recover a particular result. The solution to this problem is an effect known as *interference*. It is sometimes possible to force unwanted states to destructively interfere with each another. In this way, the unwanted results can be removed, leaving the single, desired outcome behind.

Quantum computers are well suited to puzzles that require many alternatives to be evaluated and a single result returned. Combinatorial optimization problems match this requirement rather nicely (see Chapter 6). The Travelling Salesman Problem, for example, requires that the length of all possible city tours be evaluated and only the shortest returned. This fits perfectly with the architecture of a quantum computer, provided that the suboptimal solutions can be made to interfere. For combinatorial optimization problems, quantum computers promise performance far in excess of the world's fastest supercomputers. A working quantum computer would revolutionize challenging problems such as drug discovery, materials design, and scheduling. It would also crack the prime factorization problem.

Shor's algorithm for finding prime factors is slow on a conventional computer but is well-suited to a quantum computer. The algorithm begins by guessing one of the primes. Of course, this guess is almost certainly incorrect. Rather than guess again, Shor's algorithm seeks to improve this guess. It does this by multiplying the guess by itself over and over again. Every time, it divides the original large number by the result of the multiplication and stores the remainder. After a large number of repetitions, the sequence of remainder values displays a pattern—the series repeats with a fixed period (see clock arithmetic in Chapter 7). Shor's algorithm determines this period by means of the Fourier transform (see Chapter 2). The peak of the Fourier transform output identifies the sequence's period. A multiple of the sought-after prime can be calculated as the original guess to the power of the period divided by two, all minus one.

At this point, the algorithm has the original large number and a multiple of one of its prime factors. These two numbers are both multiples of the desired prime. Finding the greatest common divisor of two numbers is relatively straight-forward. Euclid's algorithm can be applied (see Chapter 1). Euclid's algorithm repeatedly subtracts one number from the other until the working values are equal. When this happens, the working values are equal to the greatest common divisor. In Shor's algorithm, the greatest common divisor is one of the prime factors. The other prime factor can be found simply by dividing the original large number by this prime.

The procedure doesn't work every time. Success depends on the initial guess. If the routine fails, the steps are repeated starting with a different guess. In ninety-nine per cent of cases, Shor's algorithm produces the prime factors in ten or fewer iterations (see the Appendix for more details).

On a conventional computer, multiplying the guess by itself over and over again is very slow. The loop has to be repeated a large number of times before any pattern appears. On a quantum computer, these multiplications can be performed simultaneously, thanks to superposition. After that, a quantum Fourier transform can be used to cancel all but the strongest repeating pattern. This gives the period of the remainder sequence, which can be collapsed and measured. Euclid's algorithm is then performed on a conventional computer. Superposition and interference allow the quantum computer to perform Shor's algorithm amazingly quickly.

Teams at Google, IBM, Microsoft, and a handful of start-ups are now chasing the quantum computing dream. Their devices bear greater resemblance to large physics experiments than supercomputers. Building a quantum computer requires design and subatomic fabrication of quantum logic gates. Measuring and controlling the state of the subatomic particles requires incredibly precise equipment. To perform reliable measurements, the qubits must be cooled to near absolute zero ($-273°C$).

To date, computation with up to seventy-two qubits has been demonstrated. In theory, seventy-two qubits should provide immense computing power. However, in practice, *quantum noise* affects performance. Minute fluctuations in the state of the subatomic particles can lead to errors in computation. Teams compensate for this by dedicating some of the qubits to error correction (see Chapter 7). The downside is that fewer qubits are available for computation. On the face of it, the solution appears to be straightforward—simply add more qubits. However, there is a worry. What if more qubits just mean more noise and errors? What if none are available for computation?

In October 2019, a team from Google claimed that their quantum computer had attained *quantum supremacy*. The group stated that the computer had performed a computation that could not conceivably be completed on a conventional computer. The program checked that the output of a quantum random number generator was truly random. Their Sycamore quantum computing chip completed the task in 200 seconds using fifty-three qubits. The team estimated that the same calculation would take more than 10,000 years on a supercomputer. IBM begged to differ. They calculated that the task could be performed in two and a half days on a supercomputer. Not quantum supremacy then, but, still, the difference between 4 and 3,600 minutes is stark.

A great many challenges remain. However, it does appear that the quantum computer designers are on to something big.

Not The End

Algorithms have come a long way since they were first etched into clay tablets in ancient Mesopotamia. The first computers transformed their importance and capability. Since the invention of the integrated circuit, the power of algorithms has risen exponentially. It may be that further acceleration will come from quantum computing. It is difficult to see more than a few years ahead, but it seems that AI appears set to radically change the way that our world works.

A dearth of translators has meant that thousands of clay tablets from ancient Mesopotamia lie unread in the museums of the world. Today, the latest AI algorithms are being put to the task of autonomously translating 67,000 administrative tablets from twenty-first century BCE southern Mesopotamia. In perhaps the longest of full circles, the newest algorithms are about to interpret the oldest.

Appendix

PageRank Algorithm

Take the table of link counts as input.
Calculate the PageRanks as the number of incoming links for a page
    divided by the average number of incoming links.
Repeat the following:
    Repeat the following for every column:
        Set a running total to zero.
        Repeat the following for every entry in the column:
            Look up the current PageRank for the row.
            Multiply by the number of links between the row and
                column.
            Divide by the total number of outgoing links for the row.
            Multiply by the damping factor.
            Add to the running total.
        Stop repeating when all entries in the column have been
            processed.
        Add the damping term to the running total.
        Store this value as the new PageRank for the column.
    Stop repeating when all columns have been processed.
Stop repeating when the change in the PageRanks is small.
Output the PageRanks.

Artificial Neural Network Training

Take the training dataset and network topology as input.
Populate the topology with random parameters.
Repeat the following:
    Repeat for every training example:
        Apply the input to the network.
        Calculate the network output using forward propagation.
        Calculate the error between the actual and desired outputs.
        Repeat for every layer moving backwards through the
            network:
            Repeat for every neuron in the layer:
                Repeat for every weight and bias in the neuron:
                    Determine the relationship between the
                        parameter and the error.
                    Calculate a correction value for the parameter.
                    Multiply the correction by the training rate.
                    Subtract this value from the parameter.
                Stop repeating when the neuron has been updated.
            Stop repeating when the layer has been updated.
        Stop repeating when network has been updated.
    Stop repeating when the training dataset has been exhausted.
Stop repeating when there is no further reduction in the error.
Freeze the parameters.
Training is complete.

Bitcoin Algorithm

The bitcoin sender:
>   Creates a transaction recording the sender's public key, the
>   receiver's public key, the amount, and the IDs of the inputs to
>   the transaction.
>
>   Appends a digital signature to the transaction.
>
>   Broadcasts the signed transaction to the Bitcoin network.

The computers on the Bitcoin network:
>   Check that the signature is authentic.
>
>   Check that the input transactions have not been spent.
>
>   Incorporate the transaction in a candidate block.
>
>   Link the candidate to the chain.
>
>   Repeat the following steps:
>   >   Generate a random number and append it to the block.
>   >   Calculate the hash for the block.
>
>   Stop repeating when the hash is less than the threshold or
>   abandon the hunt when another computer wins the race.
>
>   Broadcast the valid block to the network.

The bitcoin receiver:
>   Accepts the transaction when it and five more blocks have been
>   added to the chain.

Shor's Algorithm

Take a large number as input.
Repeat the following steps:
> Take a prime number as a guess.
> Store the guess in memory.
> Create an empty list.
> Repeat the following steps:
>> Multiplying the value in memory by the guess.
>> Update the value in memory.
>> Calculate the remainder after dividing the input by the value in memory.
>> Append this remainder to the list.
>
> Stop repeating after a large number of repetitions.
> Apply the Fourier transform to the list of remainders.
> Identify the period of the strongest harmonic.
> Calculate the guess to the power of the period divided by two, all minus one.
> Apply Euclid's algorithm to this value and the input.

Stop repeating when the value returned is a prime factor of the input.
Divide the input by the prime factor.
Output both prime factors.

Notes

Introduction

(Page 1) Strictly speaking, even addition is an algorithm.

(Page 1) A common misconception is that the word 'algorithm' is synonymous with 'method'. The two words are not equivalent. A method is a series of steps. An algorithm is a series of steps that solve an information problem.

(Page 3) In this example, the books are considered as symbols representing the titles of the books. Rearranging the books—the symbols—has the effect of sorting the titles.

Chapter 1 Ancient Algorithms

(Page 10) The Incas were the only Bronze age civilization that, as far as we know, did not invent writing. Egyptian mathematics was recorded on papyrus. As result, it is suspected that much has been lost. Ancient Egyptian mathematics was practical in nature and revolved around numerical calculation. Mesopotamian mathematics was more explicit in the usage, application, and description of algorithms.

(Page 11) Originally a city state in central Mesopotamia, the Akkadian Empire grew to encompass the bulk of the land between the rivers and parts of the Levant.

(Page 16) Heron's algorithm is a simplification of the more general Newton–Raphson method.

(Page 16) The approximation algorithm can be accelerated by dividing two by the most recent approximation, rather than taking the value halfway between the most recent approximations.

(Page 17) In 1994, Jerry Bonnell and Robert Nemiroff programmed a VAX computer to enumerate the square root of two to ten million digits. Bonnell and Nemiroff did not disclose which algorithm they used.

(Page 20) The original version of Euclid's algorithm used subtraction. It is possible to use division instead. In some cases, using division is faster, but it must be remembered that a single division operation is, in fact, a series of subtraction operations. Alternatively, division can be performed as subtraction in the log domain.

Chapter 2 Ever-Expanding Circles

(Page 27) It has been claimed that Archimedes' screw was invented by the Babylonians and the design transported to Egypt in the time of Archimedes.

(Page 28) Archimedes did not have the benefit of the sine, cosine, and tangent trigonometric functions that we use today. The length of a side of the inner hexagon is $2r\sin(\frac{\pi}{6})$. The angle is the angle from the centre to the bisection of the side. The length of a side of the outer hexagon is $2r\tan(\frac{\pi}{6})$.

(Page 29) Archimedes' algorithm was finally supplanted by calculations based on infinite series.

(Page 30) Quadratic algorithms are of the form $ax^2 + bx + c = 0$ where a, b, and c are known constants, or coefficients, and x is the unknown value to be determined.

(Page 30) *The Compendious Book on Calculation by Completion and Balancing* was translated into Latin by Robert of Chester around 1145.

(Page 31) Other cultures developed decimal numbers systems, including the Chinese and Egyptians. However, they used different numerals (digit representations) and, by and large, used alternative positional systems.

(Page 33) More precisely, Fourier claimed that any function of a variable could be expressed as the summation of a series of sinusoidal functions whose periods are power of two divisors of the period of the original function. The Fourier series had been previously used by Leonhard Euler, Joseph Louis Lagrange, and Carl Friedrich Gauss. However, Fourier's work served to popularize the concept and was the basis of later work.

(Page 35) In the Fourier transform example, I omit the DC (constant) component for simplicity.

(Page 37) Tukey also has the distinction of coining the terms 'software' and 'bit'.

Chapter 3 Computer Dreams

(Page 42) The finished section of the Difference Engine No. 1 is now on display in the Science Museum in London. The unit is cuboid—just over 60 cm high, 60 cm wide and almost 45 cm in length. A wooden base supports a metal framework containing three stacks of brass discs. The discs are labelled with the decimal digits and are interconnected by an intricate mechanism of shafts, levers and gears. A crank and a series of cogwheels sit atop the engine, above a metal plate. The quality and precision of Clement's workmanship is evident. However, to the modern eye, the device looks more like an ingenious, Victorian cash register than the makings of a computer.

(Page 45) Lovelace's participation in the Analytic Engine project has been overstated in some quarters. She was not involved in the design of the machine itself. However, she did understand what it did, how it could be used for computation, and how to program it. She challenged Babbage and pushed him to explain his methods. Not only did she consider what the machine was, but also she envisioned what it might become. Perhaps her greatest achievement lay in communicating Babbage's extraordinary ideas to a wider audience.

(Page 45) A section of the Analytic Engine was assembled and is now in the Science Museum, London.

(Page 45) Macabrely, half of Babbage's brain is on display in the Science Museum, London. The other half is held in the Royal College of Surgeons. Menabrea, author of the original paper on the Analytic Engine, went on to become Prime Minister of Italy (1867–1869).

(Page 50) Turing's supervisor at Princeton, Alonzo Church, came up with an alternative, calculus-based proof at roughly the same time. Turing's proposal was closely related to earlier work by Kurt Gödel.

(Page 52) Turing's original description of the Turing Test, rather oddly, equates differentiating between a computer and human with differentiating between a man and woman. One wonders if there was a subtext regarding his own homosexuality.

(Page 53) It has been reported that the Apple logo was emblematic of the apple found by Turing's bedside. When asked, Steve Jobs replied that it wasn't, but he wished it had been.

(Page 54) A hack to make the Z3 Turing Complete was published in 1998.

(Page 54) Under the direction of George Stibitz, Bell Labs also developed a relay-based calculator.

Chapter 4 Weather Forecasts

(Page 59) Herein, I use the word 'computer' as shorthand for 'pseudo Turing Complete computer'. Pseudo in that they do not have infinite memory. Pseudo Turing Complete computers are digital in that they process numbers and these numbers represent information. So-called 'analog computers' are fixed-function devices that use a continuous physical quantity to represent information.

(Page 63) A key part of Larson's ruling was 'Eckert and Mauchly did not themselves first invent the automatic electronic digital computer, but instead derived that subject matter from one Dr. John Vincent Atanasoff.'

(Page 66) The law of large numbers states that the average outcome of a number of trials of a random process tends to the true value as more trials as performed.

(Page 67) Enrico Fermi previously experimented with a version of the Monte Carlo method but did not publish on the topic.

(Page 67) Metropolis later lived up to his box office name by appearing as a scientist in a Woody Allen movie.

(Page 71) Henri Poincaré identified a chaotic system in the 1880s in the form of three bodies orbiting one another. He also developed theory to investigate the effect.

(Page 73) Today, the number of transistors in an integrated circuit doubles every twenty-four months—a slight deceleration.

Chapter 5 Artificial Intelligence Emerges

(Page 77) The British Broadcasting Corporation recorded three of Strachey's computer compositions: *The National Anthem*, *Baa Baa Black Sheep*, and *In The Mood*. Restorations of the historic recordings are now online.

(Page 86) In 1971, Shaw left RAND to work as a software and programming consultant. He passed away in 1991.

(Page 90) Chinook, a Checkers-playing program written by Jonathan Schaeffer, defeated world champion Marion Tinsley in 1994.

(Page 91) Samuel's learning and minimax procedures drew on suggestions made by Claude Shannon in a 1950 paper regarding Chess. Unlike Samuel, Shannon did not develop an actual program.

(Page 91) Newell and Simon did make three other predictions that did come true.

Chapter 6 Needles in Haystacks

(Page 96) Currently, the fastest algorithm for solving the Travelling Salesman Problem has exponential complexity. In 1976, Nicos Christofides came up with an algorithm that quickly produces routes that are guaranteed to be at most fifty percent worse than the minimum route. Since then, fast approximation algorithms have been improved so as to give a guarantee of forty percent of the minimum.

(Page 98) In the worst case, Quicksort takes as many operations as Insertion Sort.

(Page 102) The only Millennium Problem to have been solved so far is the Poincaré Conjecture by Grigori Perelman in 2003.

(Page 103) George Forsythe has been credited with coining the term 'computer science' in a paper published in 1961, but the term is older than that. Luis Fein used it in a paper in 1959 to describe university computing schools.

(Page 110) The original NRMP algorithm was developed by John Mullin and J.M. Stalnaker, prior to the adoption of the Boston Pool algorithm. The Gale–Shapley algorithm was rejected twice for being too simple before it was finally published in 1962.

(Page 110) Prior to Holland's work, Nils Barricelli and Alexander Fraser used computer algorithms to model and study biological evolutionary processes. However, their proposals lacked certain key elements contained in Holland's work.

(Page 111) It has been claimed that Holland was the first person to receive a PhD in Computer Science in the US. In fact, he was enrolled to a Communication Sciences graduate programme at the University of Michigan, not a Computer Science programme. The first two PhD degrees in Computer Science in the

US were awarded on the same day—7 June 1965. The recipients were Sister May Kemmar at the University of Wisconsin and Irving Tang at Washington University in St. Louis.

(Page 111) Fisher dedicated his book to Darwin's son, Leonard Darwin, with whom Fisher had a long friendship and who provided much support in the writing of the book.

(Page 113) Paradoxically, Holland used the success of natural evolution to justify his work on genetic algorithms, whereas biologists employed Holland's algorithms to support their arguments for the existence of natural evolution.

Chapter 7 The Internet

(Page 120) There is some disagreement about the role of Leonard Kleinrock in the development of packet-switching. In my view, during his PhD at MIT, Kleinrock developed mathematical analyses that were applicable to packet-switched networks, but he did not invent packet-switching.

(Page 126) The word 'Internet', a contraction of internetworking, seems to have been coined by Vint Cerf and two colleagues at Stanford—Yogen Dalal and Carl Sunshine.

(Page 129) The ISBN check digit is calculated from the other twelve digits by multiplying by 1 or 3 alternating, adding the results together, extracting the last decimal digit from the sum, subtracting the digit from 10, and, if necessary, replacing a resulting 10 with a 0. For example, the novel *A Game of Thrones* has the ISBN 978-000754823-1. The checksum is $(9 \times 1) + (3 \times 7) + (8 \times 1) + (3 \times 0) + (0 \times 1) + (3 \times 0) + (7 \times 1) + (3 \times 5) + (4 \times 1) + (3 \times 8) + (2 \times 1) + (3 \times 3) = 99$. The check digit is then $10 - 9 = 1$. There is a nine in ten chance that a single digit transcription error will be detected by cross-validation of the ISBN check digit.

(Page 132) The quick way to determine the bit position in-error is to write the parity check results down in reverse order. In the example, this gives 0011 where 0 indicates an even count (no error in that group) and 1 indicates an odd count (an error in that group). This value can be interpreted as a binary number 0011 giving the position of the error, in this case, position 3.

(Page 135) Merkle followed up with a paper outlining his own ideas in 1978.

(Page 136) The Alice, Bob, and Eve (eavesdropper) characters were invented by Ron Rivest, Adi Shamir, and Leonard Adleman to explain their new encryption algorithm. The characters have taken on a life of their own and are now routinely referred to in papers on cryptography and security.

(Page 139) Formally, the totient is the number of integers less than a number that are co-prime to it, i.e. they share no factors. The public exponent is a number between one and the totient where the chosen number and the totient are co-prime. Co-prime means that they must not be both evenly divisible by the same number, other than one. A simple solution is just to choose the public exponent as a prime number less than the totient.

Chapter 8 Googling the Web

(Page 146) Mosaic was soon displaced by Netscape Navigator. Microsoft licensed Mosaic for the development of Internet Explorer.

Chapter 9 Facebook and Friends

(Page 166) In truth, a lot of other factors can be used. For example, good recommenders don't just select similar users and similar movies. All users and movies can be used as predictors. Imagine that Ken and Jill never agree on movies. Let's say Ken's ratings are always the exact opposite of Jill's. If Ken says 1 star, then Jill says 5 stars, and so on. Even though their scores are dissimilar, Ken's ratings are, in fact, the perfect predictor of Jill's. Just subtract Ken's score from 6. The fact that their ratings histories are always dissimilar is helpful information.

Chapter 10 America's Favourite Quiz Show

(Page 172) I don't discuss the Deep Blue versus Kasparov match in the book because it is more a computer chip design story than an algorithm story.
(Page 172) Gary Tesauro of TD-Gammon fame worked on the game playing strategy elements of the system.

Chapter 11 Mimicking the Brain

(Page 186) When counting the number of layers in a network, the input layer is excluded.
(Page 188) Minsky and Rosenblatt both attended the Bronx High School of Science.
(Page 188) I have seen claims of forthright face-to-face debates between Minsky and Rosenblatt but I have not come across any first-hand accounts of same.
(Page 195) Hinton is a great-great-grandson of George Boole.
(Page 197) The term Deep Learning was coined by Rina Dechter in 1986 in reference to machine learning and in 2000 by Igor Aizenberg in reference to neural networks.
(Page 201) Google took over funding of the Turing Award in 2014, quadrupling the prize fund to one million dollars.

Chapter 13 Next Steps

(Page 217) The smallest Bitcoin unit is the satoshi—one hundred millionth of a bitcoin.

(Page 217) The word for the Bitcoin concept is in title case, whereas references to bitcoins are in lower case.

(Page 224) In the event that RSA is broken, Bitcoin can switch to postquantum cryptographical techniques such as elliptic curve cryptography.

(Page 224) Feynman's 1985 best-selling autobiography—*Surely You're Joking, Mr. Feynman!*—made him truly famous.

Permissions

Bibliography

1. Hoare, C.A.R., 1962. Quicksort. *The Computer Journal*, 5(1), pp. 10–16.
2. Dalley, S., 1989. *Myths from Mesopotamia*. Oxford: Oxford University Press.
3. Finkel, I., 2014. *The Ark before Noah*. Hachette.
4. Rawlinson, H.C., 1846. The Persian cuneiform inscription at Behistun, decyphered and translated. *Journal of the Royal Asiatic Society of Great Britain and Ireland*, 10, pp. i–349.
5. Knuth, D.E., 1972. Ancient Babylonian algorithms. *Communications of the ACM*, 15(7), pp. 671–7.
6. Fowler, D. and Robson, E., 1998. Square root approximations in old Babylonian mathematics: YBC 7289 in context. *Historia Mathematica*, 25(4), pp. 366–78.
7. Fee, G.J., 1996. The square root of 2 to 10 million digits. http://www.plouffe.fr/simon/constants/sqrt2.txt. (Accessed 5 July 2019).
8. Harper, R.F., 1904. *The Code of Hammurabi, King of Babylon*. Chicago: The University of Chicago Press.
9. Jaynes, J., 1976. *The Origin of Consciousness in the Breakdown of the Bicameral Mind*. New York: Houghton Mifflin Harcourt.
10. Boyer, C.B. and Merzbach. U.C., 2011. *A History of Mathematics*. Oxford: John Wiley & Sons.
11. Davis, W.S., 1913. *Readings in Ancient History, Illustrative Extracts from the Source: Greece and the East*. New York: Allyn and Bacon.
12. Beckmann, P., 1971. *A history of Pi*. Boulder, CO: The Golem Press.
13. Mackay, J.S., 1884. Mnemonics for π, $\frac{1}{\pi}$, e. *Proceedings of the Edinburgh Mathematical Society*, 3, pp. 103–7.
14. Dietrich, L., Dietrich, O., and Notroff, J., 2017. Cult as a driving force of human history. *Expedition Magazine*, 59(3), pp. 10–25.
15. Katz, V., 2008. *A History of Mathematics*. London: Pearson.
16. Katz, V. J. ed., 2007. *The Mathematics of Egypt, Mesopotamia, China, India, and Islam*. Princeton, NJ: Princeton University Press.
17. Palmer, J., 2010. Pi record smashed as team finds two-quadrillionth digit – BBC News [online]. https://www.bbc.com/news/technology-11313194, September 16 2010. (Accessed 6 January 2019).
18. The Editors of Encyclopaedia Britannica, 2020. Al-Khwārizmī. In Encyclopaedia Britannica [online]. https://www.britannica.com/biography/al-Khwarizmi (Accessed 20 May 2020).
19. LeVeque, W.J. and Smith, D.E., 2019. Numerals and numeral systems. In Encyclopaedia Britannica [online]. https://www.britannica.com/science/numeral. (Accessed 19 May 2020).
20. The Editors of Encyclopaedia Britannica, 2020. French revolution. In Encyclopaedia Britannica [online]. https://www.britannica.com/event/French-Revolution. (Accessed 19 May 2020).

21. Cooley, J.W. and Tukey, J.W., 1965. An algorithm for the machine calculation of complex Fourier series. *Mathematics of Computation*, 19(90), pp. 297–301.
22. Rockmore, D.N., 2000. The FFT: An algorithm the whole family can use. *Computing in Science & Engineering*, 2(1), pp. 60–4.
23. Anonymous, 2016. James William Cooley. *New York Times*.
24. Heidenman, C., Johnson, D., and Burrus, C., 1984. Gauss and the history of the fast Fourier transform. *IEEE ASSP Magazine*, 1(4), pp. 14–21.
25. Huxley, T.H., 1887. *The Advance of Science in the Last Half-Century*. New York: Appleton and Company.
26. Swade, D., 2000. *The Cogwheel Brain*. London: Little, Brown.
27. Babbage, C., 2011. *Passages from the Life of a Philosopher*. Cambridge: Cambridge University Press.
28. Menabrea, L.F. and King, A., Countess of Lovelace, 1843. Sketch of the analytical engine invented by Charles Babbage. *Scientific Memoirs*, 3, pp. 666–731.
29. Essinger, J., 2014. *Ada's algorithm: How Lord Byron's daughter Ada Lovelace Launched the Digital Age*. London: Melville House.
30. Kim, E.E. and Toole, B.A., 1999. Ada and the first computer. *Scientific American*, 280(5), pp. 76–81.
31. Isaacson, W., 2014. *The Innovators*. New York: Simon and Schuster.
32. Turing, S., 1959. *Alan M. Turing*. Cambridge: W. Heffer & Sons, Ltd.
33. Turing, A.M., 1937. On computable numbers, with an application to the Entscheidungsproblem. *Proceedings of the London Mathematical Society*, s2–42(1), pp. 230–65.
34. Davis, M., 1983. *Computability and Unsolvability*. Mineola, NY: Dover Publications.
35. Strachey, C., 1965. An impossible program. *The Computer Journal*, 7(4), p. 313.
36. Turing, A.M., 1950. Computing machinery and intelligence. *Mind*, 59(236), pp. 433–60.
37. Copeland, B. Jack., 2014. *Turing*. Oxford: Oxford University Press.
38. Abbe, C., 1901. The physical basis of long-range weather forecasts. *Monthly Weather Review*, 29(12), pp. 551–61.
39. Lynch, P., 2008. The origins of computer weather prediction and climate modeling. *Journal of Computational Physics*, 227(7), pp. 3431–44.
40. Hunt, J.C.R., 1998. Lewis Fry Richardson and his contributions to mathematics, meteorology, and models of conflict. *Annual Review of Fluid Mechanics*, 30(1), pp. xiii–xxxvi.
41. Mauchly, J.W., 1982. The use of high speed vacuum tube devices for calculating. In: B. Randall, ed., *The Origins of Digital Computers*. Berlin: Springer, pp. 329–33.

42. Fritz, W.B., 1996. The women of ENIAC. *IEEE Annals of the History of Computing*, 18(3), pp. 13–28.
43. Ulam, S., 1958. John von Neumann 1903–1957. *Bulletin of the American Mathematical Society*, 64(3), pp. 1–49.
44. Poundstone, W., 1992. *Prisoner's Dilemma*. New York: Doubleday.
45. McCorduck, P., 2004. *Machines Who Think*. Natick, MA: AK Peters.
46. Goldstein, H.H., 1980. *The Computer from Pascal to von Neumann*. Princeton, NJ: Princeton University Press.
47. Stern, N., 1977. *An Interview with J. Presper Eckert*. Charles Babbage Institute, University of Minnesota.
48. Von Neumann, J., 1993. First draft of a report on the EDVAC. *IEEE Annals of the History of Computing*, 15(4), pp. 27–75.
49. Augarten, S., 1984. A. W. Burks, 'Who invented the general-purpose electronic computer?' In *Bit by bit: An Illustrated History of Computers*. New York: Ticknor & Fields. Epigraph, Ch. 4.
50. Kleiman, K., 2014. The computers: The remarkable story of the ENIAC programmers. Vimeo [online]. https://vimeo.com/ondemand/eniac6. (Accessed 11 March 2019).
51. Martin, C.D., 1995. ENIAC: Press conference that shook the world. *IEEE Technology and Society Magazine*, 14(4), pp. 3–10.
52. Nicholas Metropolis. The beginning of the Monte Carlo method. *Los Alamos Science*, 15(584), pp. 125–30.
53. Eckhardt, R., 1987. Stan Ulam, John von Neumann, and the Monte Carlo method. *Los Alamos Science*, 15(131–136), p. 30.
54. Wolter, J., 2013. Experimental analysis of Canfield solitaire. http://politaire.com/article/canfield.html. (Accessed 20 May 2020).
55. Metropolis, N. and Ulam, S., 1949. The Monte Carlo method. *Journal of the American Statistical Association*, 44(247), pp. 335–341.
56. Charney, J.G. and Eliassen, A., 1949. A numerical method for predicting the perturbations of the middle latitude westerlies. *Tellus*, 1(2), pp. 38–54.
57. Charney, J.G., 1949. On a physical basis for numerical prediction of large-scale motions in the atmosphere. *Journal of Meteorology*, 6(6), pp. 372–85.
58. Platzman, G.W., 1979. The ENIAC computations of 1950: Gateway to numerical weather prediction. *Bulletin of the American Meteorological Society*, 60(4), pp. 302–12.
59. Charney, J.G., Fjörtoft, R., and von Neumann, J., 1950. Numerical integration of the barotropic vorticity equation. *Tellus*, 2(4), pp. 237–54.
60. Blair, C., 1957. Passing of a great mind. *Life Magazine*, 42(8), pp. 89–104.
61. Lorenz, E.N., 1995. *The Essence of Chaos*. Seattle: University of Washington Press.
62. Lorenz, E.N., 1963. Deterministic nonperiodic flow. *Journal of the Atmospheric Sciences*, 20(2), pp. 130–41.

63. Epstein, E.S., 1969. Stochastic dynamic prediction. *Tellus*, 21(6), pp. 739–59.

64. European Centre for Medium-Range Weather Forecasts, 2020. Advancing global NWP through international collaboration. http://www.ecmwf.int. (Accessed 19 May 2020).

65. Lynch, P. and Lynch, O., 2008. Forecasts by PHONIAC. *Weather*, 63(11), pp. 324–6.

66. Shannon, C.E., 1950. Programming a computer for playing chess. *Philosophical Magazine*, 41(314), pp. 256–75.

67. National Physical Laboratory, 2012. Piloting Computing: Alan Turing's Automatic Computing Engine. YouTube [online]. https://www.youtube.com/watch?v=cEQ6cnwaY_s. (Accessed 27 October 2019).

68. Campbell-Kelly, M., 1985. Christopher Strachey, 1916–1975: A biographical note. *Annals of the History of Computing*, 7(1), pp. 19–42.

69. Copeland, J. and Long, J., 2016. Restoring the first recording of computer music. https://blogs.bl.uk/sound-and-vision/2016/09/restoring-the-first-recording-of-computer-music.html#. (Accessed 15 February 2019).

70. Foy, N., 1974. The word games of the night bird (interview with Christopher Strachey). *Computing Europe*, 15, pp. 10–11.

71. Roberts, S., 2017. Christopher Strachey's nineteen-fifties love machine. *The New Yorker*, February 14.

72. Strachey, C., 1954. The 'thinking' machine. *Encounter*, III, October.

73. Strachey, C.S., 1952. Logical or non-mathematical programmes. In *Proceedings of the 1952 ACM National Meeting* New York: ACM. pp. 46–9.

74. McCarthy, J., Minsky, M.L., Rochester, N., and Shannon, C.E., 2006. A proposal for the Dartmouth summer research project on artificial intelligence, August 31, 1955. *AI Magazine*, 27(4), pp. 12–14.

75. Newell, A. and Simon, H., 1956. The logic theory machine: A complex information processing system. *IRE Transactions on Information Theory*. 2(3), pp. 61–79.

76. Newell, A., Shaw, J.C., and Simon, H.A., 1959. Report on a general problem solving program. In *Proceedings of the International Conference on Information Processing*. Paris: UNESCO. pp. 256–64.

77. Newell, A. and Simon, H., 1972. *Human Problem Solving*. New York: Prentice-Hall.

78. Schaeffer, J., 2008. *One Jump Ahead: Computer Perfection at Checkers*. New York: Springer.

79. Samuel, A.L., 1959. Some studies in machine learning using the game of checkers. *IBM Journal of Research and Development*, 3(3), pp. 210–29.

80. McCarthy, J. and Feigenbaum, E.A., 1990. In memoriam: Arthur Samuel: Pioneer in machine learning. *AI Magazine*, 11(3), p. 10.

81. Samuel, A.L., 1967. Some studies in machine learning using the game of checkers. ii. *IBM Journal of Research and Development*, 11(6), pp. 601–17.

82. Madrigal, A.C., 2017. How checkers was solved. *The Atlantic*. July 19.

83. Simon, H.A., 1998. Allen Newell: 1927–1992. *IEEE Annals of the History of Computing*, 20(2), pp. 63–76.
84. CBS, 1961. The thinking machine. YouTube [online]. https://youtu.be/aygSMgK3BEM. (Accessed 19 May 2020).
85. Dreyfus, H.L., 2005. Overcoming the myth of the mental: How philosophers can profit from the phenomenology of everyday expertise. In: *Proceedings and Addresses of the American Philosophical Association*, 79(2), pp. 47–65.
86. Nilsson, N.J., 2009. *The Quest for Artificial Intelligence*. Cambridge: Cambridge University Press.
87. Schrijver, A., 2005. On the history of combinatorial optimization (till 1960). In: K. Aardal, G.L. Nemhauser, R. Weismantel, eds., *Discrete optimization*, vol. 12. Amsterdam: Elsevier. pp. 1–68.
88. Dantzig, G., Fulkerson, R., and Johnson, S., 1954. Solution of a large-scale traveling-salesman problem. *Journal of the Operations Research Society of America*, 2(4), pp. 393–410.
89. Cook, W., n.d. Traveling salesman problem. http://www.math.uwaterloo.ca/tsp/index.html. (Accessed 19 May 2020).
90. Cook, S.A., 1971. The complexity of theorem-proving procedures. In: *Proceedings of the 3rd annual ACM Symposium on Theory of Computing*. New York: ACM. pp. 151–8.
91. Karp, R., n.d. A personnal view of Computer Science at Berkeley. https://www2.eecs.berkeley.edu/bears/CS_Anniversary/karp-talk.html. (Accessed 15 February 2019).
92. Garey, M.R. and Johnson, D.S., 1979. *Computers and Intractability*. New York: W. H. Freeman and Company.
93. Dijkstra, E.W., 1972. The humble programmer. *Communications of the ACM*, 15(10), pp. 859–66.
94. Dijkstra, E.W., 2001. Oral history interview with Edsger W. Dijkstra. Technical report, Charles Babbage Institute, August 2.
95. Dijkstra, E.W., 1959. A note on two problems in connexion with graphs. *Numerische mathematik*, 1(1), pp. 269–71.
96. Darrach, B., 1970. Meet Shaky: The first electronic person. *Life Magazine*, 69(21):58B–68B.
97. Hart, P.E., Nilsson, N.J., and Raphael, B., 1968. A formal basis for the heuristic determination of minimum cost paths. *IEEE Transactions on Systems Science and Cybernetics*, 4(2), pp. 100–7.
98. Hitsch, G.J., Hortaçsu, A., and Ariely, D., 2010. Matching and sorting in online dating. *The American Economic Review*, 100(1), pp. 130–63.
99. NRMP. National resident matching program. http://www.nrmp.org. (Accessed 19 May 2020).
100. Roth, A.E., 2003. The origins, history, and design of the resident match. *Journal of the American Medical Association*, 289(7), pp. 909–12.

101. Roth, A.E., 1984. The evolution of the labor market for medical interns and residents: A case study in game theory. *The Journal of Political Economy*, 92, pp. 991–1016.

102. Anonymous, 2012. Stable matching: Theory, evidence, and practical design. Technical report, The Royal Swedish Academy of Sciences.

103. Kelly, K., 1994. *Out of Control*. London: Fourth Estate.

104. Vasbinder, J.W., 2014. *Aha... That is Interesting!: John H. Holland, 85 years young*. Singapore: World Scientific.

105. London, R.L., 2013. Who earned first computer science Ph.D.? *Communications of the ACM : blog@CACM*, January.

106. Scott, N.R., 1996. The early years through the 1960's: Computing at the CSE@50. Technical report, University of Michigan.

107. Fisher, R.A., 1999. *The Genetical Theory of Natural Selection*. Oxford: Oxford University Press.

108. Holland, J.H., 1992. *Adaptation in Natural and Artificial Systems*. Cambridge, MA: The MIT Press.

109. Holland, J.H., 1992. Genetic algorithms. *Scientific American*, 267(1), pp. 66–73.

110. Dawkins, R., 1986. *The Blind Watchmaker*. New York: WW Norton & Company.

111. Lohn, J.D., Linden, D.S., Hornby, G.S., Kraus, W.F., 2004. Evolutionary design of an X-band antenna for NASA's space technology 5 mission. In: *Proceedings of the IEEE Antennas and Propagation Society Symposium 2004*, volume 3. Monterey, CA, 20–25 June, pp. 2313–16. New York: IEEE.

112. Grimes, W., 2015. John Henry Holland, who computerized evolution, dies at 86. *New York Times*, August 19.

113. Licklider, J.C.R., 1960. Man-computer symbiosis. *IRE Transactions on Human Factors in Electronics*, 1(1), pp. 4–11.

114. Waldrop, M.M., 2001. *The Dream Machine*. London: Viking Penguin.

115. Kita, C.I., 2003. JCR Licklider's vision for the IPTO. *IEEE Annals of the History of Computing*, 25(3), pp. 62–77.

116. Licklider, J.C.R., 1963. Memorandum for members and affiliates of the intergalactic computer network. Technical report, Advanced Research Projects Agency, April 23.

117. Licklider, J.C.R., 1965. *Libraries of the Future*. Cambridge, MA: The MIT Press.

118. Licklder, J.C.R. and Taylor, R.W., 1968. The computer as a communication device. *Science and Technology*, 76(2), pp. 1–3.

119. Markoff, J., 1999. An Internet pioneer ponders the next revolution. *The New York Times*, December 20.

120. Featherly, K., 2016. ARPANET. In Encyclopedia Brittanica. [online]. https://www.britannica.com/topic/ARPANET. (Accessed 19 May 2020).

121. Leiner, B.M., Cerf, V.G., Clark, D.D., Kahn, R.E., Kleinrock., L., Lynch, D.C., Postel, J., Robers, L.G., and Wolff, S., 2009. A brief history of the Internet. *ACM SIGCOMM Computer Communication Review*, 39(5), pp. 22–31.

122. Davies, D.W., 2001. An historical study of the beginnings of packet switching. *The Computer Journal*, 44(3), pp. 152–62.

123. Baran, P., 1964. On distributed communications networks. *IEEE Transactions on Communications Systems*, 12(1), pp. 1–9.

124. McQuillan, J., Richer, I., and Rosen, E., 1980. The new routing algorithm for the ARPANET. *IEEE Transactions on Communications*, 28(5), pp. 711–19.

125. McJones, P., 2008. Oral history of Robert (Bob) W. Taylor. Technical report, Computer History Museum.

126. Metz, C., 2012. Bob Kahn, the bread truck, and the Internet's first communion. *Wired*, August 13.

127. Vint, C. and Kahn, R., 1974. A protocol for packet network interconnection. *IEEE Transactions of Communications*, 22(5), pp. 637–48.

128. Metz, C., 2007. How a bread truck invented the Internet. The Register [online]. https://www.theregister.co.uk/2007/11/12/thirtieth_anniversary _of_first_internet_connection/. (Accessed 19 May 2020).

129. Anonymous, 2019. Number of internet users worldwide from 2005 to 2018. Statista [online]. https://www.statista.com/statistics/273018/number-of-internet-users-worldwide/. (Accessed 19 May 2020).

130. Lee, J., 1998. Richard Wesley Hamming: 1915–1998. *IEEE Annals of the History of Computing*, 20(2), pp. 60–2.

131. Suetonius, G., 2009. *Lives of the Caesars*. Oxford: Oxford University Press.

132. Singh, S., 1999. *The Code Book: The Secret History of Codes & Code-breaking*. London: Fourth Estate.

133. Diffie, W. and Hellman, M., 1976. New directions in cryptography. *IEEE Transactions on Information Theory*, 22(6), pp. 644–54.

134. Rivest, R.L., Shamir, A., Adleman, L., 1978. A method for obtaining digital signatures and public-key cryptosystems. *Communications of the ACM*, 21(2), pp. 120–6.

135. Gardner, M., 1977. New kind of cipher that would take millions of years to break. *Scientific American*. 237(August), pp. 120–4.

136. Atkins, D., Graff, M., Lenstra, A.K., Leyland, P.C., 1994. The magic words are squeamish ossifrage. In: *Proceedings of the 4th International Conference on the Theory and Applications of Cryptology*. Wollongong, Australia. November 28 November-1 December 1994. pp. 261–77. NY: Springer.

137. Levy, S., 1999. The open secret. *Wired*, 7(4).

138. Ellis, J.H., 1999. The history of non-secret encryption. *Cryptologia*, 23(3), pp. 267–73.

139. Ellis, J.H., 1970. The possibility of non-secret encryption. In: *British Communications-Electronics Security Group (CESG)* report. January.

140. Bush, V., 1945. As we may think. *The Atlantic*. 176(1), pp. 101–8.

141. Manufacturing Intellect, 2001. Jeff Bezos interview on starting Amazon. YouTube [online]. https://youtu.be/p7FgXSoqfnI. (Accessed 19 May 2020).

142. Stone, B., 2014. *The Everything Store: Jeff Bezos and the Age of Amazon*. New York: Corgi.
143. Christian, B. and Griffiths, T., 2016. *Algorithms to Live By*. New York: Macmillan.
144. Linden, G., Smith, B., and York, J., 2003. Amazon.com recommendations. *IEEE Internet Computing*, 7(1), pp. 76–80.
145. McCullough, B., 2015. Early Amazon engineer and co-developer of the recommendation engine, Greg Linden. Internet History Podcast [online]. http://www.internethistorypodcast.com/2015/04/early-amazon-engineer-and-co-developer-of-the-recommendation-engine-greg-linden/#tabpanel6. (Accessed 15 February 19).
146. MacKenzie, I., Meyer, C., and Noble, S., 2013. How retailers can keep up with consumers. Technical report, McKinsey and Company, October.
147. Anonymous, 2016. Total number of websites. Internet Live Stats [online] http://www.internetlivestats.com/total-number-of-websites/. (Accessed 15 February 19).
148. Vise, D.A., 2005. *The Google Story*. New York: Macmillian.
149. Battelle, J., 2005. The birth of Google. *Wired*, 13(8), p. 102.
150. Page, L., Brin, S., Motwani, R., and Winograd, T., 1999. The PageRank citation ranking: Bringing order to the web. Technical Report 1999–66, Stanford InfoLab, November.
151. Brin, S. and Page, L., 1998. The anatomy of a large-scale hypertextual web search engine. *Computer Networks and ISDN Systems*, 30(1–7), pp. 107–117.
152. Jones, D., 2018. How PageRank really works: Understanding Google. Majestic [blog]. https://blog.majestic.com/company/understanding-googles-algorithm-how-pagerank-works/, October 25. (Accessed 12 July 2019).
153. Willmott, D., 1999. The top 100 web sites. *PC Magazine*, February 9.
154. Krieger, L.M., 2005. Stanford earns $336 million off Google stock. *The Mercury News*, December 1.
155. Hayes, A., 2019. Dotcom bubble definition. Investopedia [online]. https://www.investopedia.com/terms/d/dotcom-bubble.asp, June 25. (Accessed 19 July 2019).
156. Smith, B. and Linden, G., 2017. Two decades of recommender systems at Amazon.com. *IEEE Internet Computing*, 21(3), pp. 12–18.
157. Debter, L., 2019. Amazon surpasses Walmart as the world's largest retailer. Forbes [online]. https://www.forbes.com/sites/laurendebter/2019/05/15/worlds-largest-retailers-2019-amazon-walmart-alibaba/#20e4cf4d4171. (Accessed 18 July 2019).
158. Anonymous, 2019. Tim Berners-Lee net worth. The Richest [online]. https://www.therichest.com/celebnetworth/celebrity-business/tech-millionaire/tim-berners-lee-net-worth/. (Accessed 22 July 2019).
159. Anonymous, 2016. Internet growth statistics. Internet World Stats [online]. http://www.internetworldstats.com/emarketing.htm. (Accessed 19 May 2020).

160. Conan Doyle, A., 1890. The sign of four. *Lippincott's Monthly Magazine*. February.
161. Kirkpatrick, D., 2010. *The Facebook Effect*. New York: Simon and Schuster.
162. Grimland, G., 2009. Facebook founder's roommate recounts creation of Internet giant. Haaretz [online]. https://www.haaretz.com/1.5050614. (Accessed 23 July 2019).
163. Kaplan, K.A., 2003. Facemash creator survives ad board. *The Harvard Crimson*, November.
164. Investors Archive, 2017. Billionaire Mark Zuckerberg: Creating Facebook and startup advice. YouTube [online]. https://youtu.be/SSly3yJ8mKU.
165. Widman, J., 2011. Presenting EdgeRank: A guide to Facebook's Newsfeed algorithm. http://edgerank.net. (Accessed 19 May 2020).
166. Anonymous 2016. Number of monthly active facebook users worldwide. Statista [online] https://www.statista.com/statistics/264810/number-of-monthly-active-facebook-users-worldwide/. (Accessed 20 May 2020).
167. Keating, G., 2012. *Netflixed*. London: Penguin.
168. Netflix, 2016. Netflix prize. http://www.netflixprize.com/. (Accessed 19 May 2020).
169. Van Buskirk, E., 2009. How the Netflix prize was won. *Wired*, September 22.
170. Thompson, C., 2008. If you liked this, you're sure to love that. *The New York Times*. November 21.
171. Piotte, M. and Chabbert, M., 2009. The pragmatic theory solution to the Netflix grand prize. Technical report, Netflix.
172. Koren, Y., 2009. The Bellkor solution to the Netflix grand prize. *Netflix Prize Documentation*, 81:1–10.
173. Johnston, C., 2012. Netflix never used its $1 million algorithm due to engineering costs. *Wired*, April 16.
174. Gomez-Uribe, C.A. and Hunt, N., 2015. The Netflix recommender system: Algorithms, business value, and innovation. *ACM Transactions on Management Information Systems*, 6(4), pp. 13:1–19.
175. Ginsberg, J., Mohebbi, M.H., Patel, R.S., Brammer, L., Smolinkski, M.S., and Brilliant, L., 2009. Detecting influenza epidemics using search engine query data. *Nature*, 457(7232), pp. 1012–14.
176. Cook, S., Conrad, C., Fowlkes, A.L., Mohebbi, M.H., 2011. Assessing Google flu trends performance in the United States during the 2009 influenza virus A (H1N1) pandemic. *PLOS ONE*, 6(8), e23610.
177. Butler, D., 2013. When Google got flu wrong. *Nature*, 494(7436), pp. 155.
178. Lazer, D., Kennedy, R., King, G., Vespignani, A., 2014. The parable of Google Flu: Traps in big data analysis. *Science*, 343(6176), pp. 1203–5.
179. Lazer, D. and Kennedy, R., 2015. What we can learn from the epic failure of Google flu trends. *Wired*, October 1.
180. Zimmer, B., 2011. Is it time to welcome our new computer overlords? *The Atlantic*, February 17.

181. Markoff, J., 2011. Computer wins on jeopardy: Trivial, it's not. *New York Times*, February 16.

182. Gondek, D.C., Lally, A., Kalyanpur, A., Murdock, J.W., Duboue, P.A., Zhang, L., Pan, Y., Qui, Z.M., and Welty, C., 2012. A framework for merging and ranking of answers in DeepQA. *IBM Journal of Research and Development*, 56(3.4), pp. 14:1–12.

183. Best, J., 2013. IBM Watson: The inside story of how the Jeopardy-winning supercomputer was born, and what it wants to do next. TechRepublic [online]. http://www.techrepublic.com/article/ibm-watson-the-inside-story-of-how-the-jeopardy-winning-supercomputer-was-born-and-what-it-wants-to-do-next/. (Accessed 15 February 2019).

184. Ferrucci, D.A., 2012. Introduction to this is Watson. *IBM Journal of Research and Development*, 56(3.4), pp. 1:1–15.

185. IBM Research, 2013. Watson and the Jeopardy! Challenge. YouTube [online]. https://www.youtube.com/watch?v=P18EdAKuC1U. (Accessed 15 September 2019).

186. Lieberman, H., 2011. Watson on Jeopardy, part 3. MIT Technology Review [online]. https://www.technologyreview.com/s/422763/watson-on-jeopardy-part-3/. (Accessed 15 September 2019).

187. Gustin, S., 2011. Behind IBM's plan to beat humans at their own game. *Wired*, February 14.

188. Lally, A., Prager, M., McCord, M.C., Boguraev, B.K., Patwardhan, S., Fan, J., Fodor, P., and Carroll J.C., 2012. Question analysis: How Watson reads a clue. *IBM Journal of Research and Development*, 56(3.4), pp. 2:1–14.

189. Fan, J., Kalyanpur, A., Condek, D.C., and Ferrucci, D.A., 2012. Automatic knowledge extraction from documents. *IBM Journal of Research and Development*, 56(3.4), pp. 5:1–10.

190. Kolodner, J.L., 1978. Memory organization for natural language data-base inquiry. Technical report, Yale University.

191. Kolodner, J.L., 1983. Maintaining organization in a dynamic long-term memory. *Cognitive Science*, 7(4), pp. 243–80.

192. Kolodner, J.L., 1983. Reconstructive memory: A computer model. *Cognitive Science*, 7(4), pp. 281–328.

193. Lohr, S., 2016. The promise of artificial intelligence unfolds in small steps. *The New York Times*, February 29. (Accessed 19 May 2020).

194. James, W., 1890. *The Principles of Psychology*. NY: Holt.

195. Hebb, D.O., 1949. *The Organization of Behavior*. NY: Wiley.

196. McCulloch, W.S. and Pitts, W., 1943. A logical calculus of the ideas immanent in nervous activity. *The Bulletin of Mathematical Biophysics*, 5(4), pp. 115–33.

197. Gefter, A., 2015. The man who tried to redeem the world with logic. *Nautilus*, February 21.

198. Whitehead, A.N. and Russell, B., 1910–1913. *Principia Mathematica*. Cambridge: Cambridge University Press.

199. Anderson, J.A. and Rosenfeld, E., 2000. *Talking Nets*. Cambridge, MA: The MIT Press.

200. Conway, F. and Siegelman, J., 2006. *Dark Hero of the Information Age: In Search of Norbert Wiener The Father of Cybernetics*. New York: Basic Books.

201. Thompson, C., 2005. Dark hero of the information age: The original computer geek. *The New York Times*, March 20.

202. Farley, B.W.A.C. and Clark, W., 1954. Simulation of self-organizing systems by digital computer. *Transactions of the IRE Professional Group on Information Theory*, 4(4), pp. 76–84.

203. Rosenblatt, F., 1958. The Perceptron: A probabilistic model for information storage and organization in the brain. *Psychological Review*, 65(6), pp. 386.

204. Rosenblatt, F., 1961. Principles of neurodynamics. perceptrons and the theory of brain mechanisms. Technical report, DTIC Document.

205. Anonymous, 1958. New Navy device learns by doing. *The New York Times*, July 8.

206. Minsky, M. and Papert, S., 1969. *Perceptrons*. Cambridge, MA: The MIT Press.

207. Minksy, M., 1952. A neural-analogue calculator based upon a probability model of reinforcement. Technical report, Harvard University Psychological Laboratories, Cambridge, Massachusetts.

208. Block, H.D., 1970. A review of Perceptrons: An introduction to computational geometry. *Information and Control*, 17(5), pp. 501–22.

209. Anonymous, 1971. Dr. Frank Rosenblatt dies at 43; taught neurobiology at Cornell. *The New York Times*, July 13.

210. Olazaran, M., 1996. A sociological study of the official history of the Perceptrons controversy. *Social Studies of Science*, 26(3), pp. 611–59.

211. Werbos, P.J., 1990. Backpropagation through time: What it does and how to do it. *Proceedings of the IEEE*, 78(10), pp. 1550–60.

212. Werbos, P.J., 1974. *Beyond regression: New tools for prediction and analysis in the behavioral sciences*. PhD. Harvard University.

213. Werbos, P.J., 1994. *The Roots of Backpropagation*, volume 1. Oxford: John Wiley & Sons.

214. Werbos, P.J., 2006. Backwards differentiation in AD and neural nets: Past links and new opportunities. In: H.M. Bücker, G. Corliss, P. Hovland, U. Naumann, and B. Norris, eds., *Automatic differentiation: Applications, theory, and implementations*. Berlin: Springer. pp. 15–34.

215. Parker, D.B., 1985. Learning-logic: Casting the cortex of the human brain in silicon. Technical Report TR-47, MIT, Cambridge, MA.

216. Lecun, Y., 1985. Une procédure d'apprentissage pour réseau a seuil asymmetrique (A learning scheme for asymmetric threshold networks). In: *Proceedings of Cognitiva 85*. Paris, France. 4–7 June, 1985. pp. 599–604.

217. Rumelhart, D.E., Hinton, G.E., and Williams, R.J., 1986. Learning representations by back-propagating errors. *Nature*, 323, pp. 533–36.
218. Hornik, K., Stinchcombe, M., and White, H., 1989. Multilayer feedforward networks are universal approximators. *Neural Networks*, 2(5), pp. 359–66.
219. Ng., A., 2018. Heroes of Deep Learning: Andrew Ng interviews Yann LeCun. YouTube [online]. https://www.youtube.com/watch?v= Svb1c6AkRzE. (Accessed 14 August 2019).
220. LeCun, Y., Boser, B., Denker, J.S., Henderson, D., Howard, R.E., Hubbard, W., and Jackel, L.D., 1989. Backpropagation applied to handwritten zip code recognition. *Neural Computation*, 1(4), pp. 541–51.
221. Thorpe, S., Fize, D., and Marlot, C., 1996. Speed of processing in the human visual system. *Nature*, 381(6582), pp. 520–2.
222. Gray, J., 2017. U of T Professor Geoffrey Hinton hailed as guru of new computing era. *The Globe and Mail*, April 7.
223. Allen, K., 2015. How a Toronto professor's research revolutionized artificial intelligence. *The Star*. April 17.
224. Hinton, G.E., Osindero, S., and Teh, Y.W., 2006. A fast learning algorithm for deep belief nets. *Neural Computation*, 18(7), pp. 1527–54.
225. Ciresan, D.C., Meier,U., Gambardella, L.M., and Schmidhuber, J., 2010. Deep big simple neural nets excel on handwritten digit recognition. *arXiv preprint arXiv:1003.0358.*
226. Jaitly, N., Nguyen, P., Senior, A.W., and Vanhoucke, V., 2012. Application of pretrained deep neural networks to large vocabulary speech recognition. In: *Proceedings of the 13th Annual Conference of the International Speech Communication Association (Interspeech)*. Portland, Oregon, 9–13 September 2012. pp. 257–81.
227. Hinton, G., et al., 2012. Deep neural networks for acoustic modeling in speech recognition: The shared views of four research groups. *IEEE Signal Processing Magazine*, 29(6), pp. 82–97.
228. Krizhevsky, A., Sutskever, Il, and Hinton, G.E., 2012. ImageNet classification with deep convolutional neural networks. In: C. Burges, ed., *Proceedings of the 27th Annual Conference on Neural Information Processing Systems 2013*. 5–10 December 2013, Lake Tahoe, NV. Red Hook, NY Curran. pp. 1097–1105.
229. Bengio, Y., Ducharme, R., Vincent, P., and Jauvin, C., 2003. A neural probabilistic language model. *Journal of Machine Learning Research*, 3, pp. 1137–55.
230. Sutskever, I., Vinyals, O., and Le, Q.V., 2014. Sequence to sequence learning with neural networks. In: Z. Ghahramani, M. Welling, C. Cortes, N.D. Lawrence, and K.Q. Weinberger, eds., *Proceedings of the 28th Annual Conference on Neural Information Processing Systems 2014*. 8–13 December 2014 , Montreal, Canada. Red Hook, NY Curran. pp. 3104–12.
231. Cho, K., Van Merriënboer, B., Bahdanau, and Bengio, Y., 2014. On the properties of neural machine translation: Encoder-decoder approaches. *arXiv preprint arXiv:1409.1259.*

232. Bahdanau, D., Cho, K, and Bengio, Y., 2014. Neural machine translation by jointly learning to align and translate. *arXiv preprint arXiv: 1409.0473.*

233. Wu, Y, et al., 2016. Google's neural machine translation system: Bridging the gap between human and machine translation. *arXiv preprint arXiv:1609.08144.*

234. Lewis-Kraus, G., 2016. The great A.I. awakening. *The New York Times,* December 20.

235. LeCun, Y, Bengio, Y., and Hinton, G., 2015. Deep learning. *Nature.* 521(7553), pp. 436–44.

236. Vincent, J., 2019. Turing Award 2018: Nobel prize of computing given to 'godfathers of AI'. The Verge [online]. https://www.theverge.com/2019/3/27/18280665/ai-godfathers-turing-award-2018-yoshua-bengio-geoffrey-hinton-yann-lecun. (Accessed 19 May 2020).

237. Foster, R.W., 2009. The classic of Go. http://idp.bl.uk/. (Accessed 20 May 2020).

238. Moyer, C., 2016. How Google's AlphaGo beat a Go world champion. *The Atlantic,* March.

239. Morris, D.Z., 2016. Google's Go computer beats top-ranked human. *Fortune,* March 12.

240. AlphaGo, 2017 [film]. Directed by Greg Kohs. USA: Reel as Dirt.

241. Wood, G., 2016. In two moves, AlphaGo and Lee Sedol redefined the future. *Wired,* March 16.

242. Metz, C., 2016. The sadness and beauty of watching Google's AI play Go. *Wired,* March 11.

243. Edwards, J., 2016. See the exact moment the world champion of Go realises DeepMind is vastly superior. Business Insider [online]. https://www.businessinsider.com/video-lee-se-dol-reaction-to-move-37-and-w102-vs-alphago-2016-3?r=US&IR=T. (Accessed 19 May 2020).

244. Silver, D., et al., 2016. Mastering the game of Go with deep neural networks and tree search. *Nature,* 529(7587), pp. 484–9.

245. Hern, A., 2016. AlphaGo: Its creator on the computer that learns by thinking. *The Guardian,* March 15.

246. Burton-Hill, C., 2016. The superhero of artificial intelligence: can this genius keep it in check? *The Guardian,* February 16.

247. Fahey, R., 2005. Elixir Studios to close following cancellation of key project. gamesindustry.biz [online]. https://www.gamesindustry.biz/articles/elixir-studios-to-close-following-cancellation-of-key-project. (Accessed 19 May 2020).

248. Mnih, V., et al., 2015. Human-level control through deep reinforcement learning. *Nature,* 518(7540), p. 529.

249. Silver, D., et al., 2017. Mastering the game of Go without human knowledge. *Nature,* 550(7676), pp. 354–9.

250. Silver, D., et al., 2018. A general reinforcement learning algorithm that masters Chess, Shogi, and Go through self-play. *Science*, 362(6419), pp. 1140–4.

251. Lovelace, B., 2018. Buffett says cryptocurrencies will almost certainly end badly. CNBC [online]. https://www.cnbc.com/2018/01/10/buffett-says-cyrptocurrencies-will-almost-certainly-end-badly.html. (Accessed 19 May 2020).

252. Anonymous, n.d. Market capitalization. blockchain.com [online]. https://www.blockchain.com/charts/market-cap. (Accessed 19 May 2020).

253. Hughes, E., 1993. A Cypherpunk's manifesto – Activism. https://www.activism.net/cypherpunk/manifesto.html. (Accessed 19 May 2020).

254. Assange, J., Appelbaum, J., Maguhn, A.M., and Zimmermann, J., 2016. *Cypherpunks: Freedom and the Future of the Internet*. London: OR books.

255. Driscoll, S., 2013. How Bitcoin works under the hood. Imponderable Things [online]. http://www.imponderablethings.com/2013/07/how-bitcoin-works-under-hood.html#more. (Accessed 19 May 2020).

256. Nakamoto, S., 2008. Bitcoin: A peer-to-peer electronic cash system. Working Paper.

257. Webster, I., 2020. Bitcoin historical prices. in2013dollars.com [online]. http://www.in2013dollars.com/bitcoin-price. (Accessed 22 June 2020).

258. Anonymous, n.d. Bitcoin all time high – how much was 1 bitcoin worth at its peak? 99BitCoins [online]. https://99bitcoins.com/bitcoin/historical-price/all-time-high/#charts. (Accessed 19 May 2020).

259. Anonymous, 2019. Bitcoin energy consumption index. https://digiconomist.net/bitcoin-energy-consumption. (Accessed 19 May 2020).

260. L.S., 2015. Who is Satoshi Nakamoto? *The Economist*, November 2.

261. Greenberg, A., 2016. How to prove you're Bitcoin creator Satoshi Nakamoto. *Wired*, April 11.

262. Feynman R.P., 1982. Simulating physics with computers. *International Journal of Theoretical Physics*, 21(6), pp. 467–88.

263. Shor, P.W., 1982. Polynomial-time algorithms for prime factorization and discrete logarithms on a quantum computer. *SIAM Review*, 41(2), pp. 303–32.

264. Anonymous, n.d. Quantum – Google AI. https://ai.google/research/teams/applied-science/quantum-ai/. (Accessed 19 May 2020).

265. Dyakonov, M., 2018. The case against quantum computing. *IEEE Spectrum*, November 15.

266. Arute, F., et al., 2019. Quantum supremacy using a programmable super-conducting processor. *Nature*, 574(7779):505–10.

267. Savage, N., 2019. Hands-on with Google's quantum computer. *Scientific American*, October 24. (Accessed 19 May 2020).

268. Anonymous, Machine translation and automated analysis of cuneiform languages (MTAAC Project). https://cdli-gh.github.io/mtaac/. (Accessed 22 June 2020).

269. Nemiroff, R. and Bonnell J., 1994. The square root of two to 10 million digits. https://apod.nasa.gov/htmltest/gifcity/sqrt2.10mil. (Accessed 19 May 2020).

270. Keough, B., 1997. Guide to the John Clifford Shaw Papers. http://sova.si.edu/record/NMAH.AC.0580. (Accessed 19 May 2020).

271. Christofides, N., 1976. Worst-case analysis of a new heuristic for the travelling salesman problem. Technical report, DTIC Document.

272. Sebo, A. and Vygen, J. 2012. Shorter tours by nicer ears. *arXiv preprint arXiv:1201.1870.*

273. Mullin, F.J. and Stalnaker, J.M., 1951. The matching plan for internship appointment. *Academic Medicine*, 26(5), pp. 341–5.

274. Anonymous, 2016. Lloyd Shapley, a Nobel laureate in economics, has died. *The Economist*, March 13.

275. Merkle, R.C., 1978. Secure communications over insecure channels. *Communications of the ACM*, 21(4), pp. 294–99.

Index